Mathias Wilhelms

Multiscale Modeling of Cardiac Electrophysiology

Adaptation to Atrial and Ventricular Rhythm Disorders and
Pharmacological Treatment

Vol. 20
Karlsruhe Transactions on Biomedical Engineering

Editor:
Karlsruhe Institute of Technology (KIT)
Institute of Biomedical Engineering

Eine Übersicht über alle bisher in dieser Schriftenreihe erschienenen Bände finden Sie am Ende des Buchs.

Multiscale Modeling of Cardiac Electrophysiology

Adaptation to Atrial and Ventricular Rhythm Disorders and Pharmacological Treatment

by
Mathias Wilhelms

Dissertation, Karlsruher Institut für Technologie (KIT)
Fakultät für Elektrotechnik und Informationstechnik, 2013
Referenten: Prof. Dr. Olaf Dössel, PD Dr. Eberhard P. Scholz,
             Prof. Dr. Uli Lemmer

Impressum

Karlsruher Institut für Technologie (KIT)
KIT Scientific Publishing
Straße am Forum 2
D-76131 Karlsruhe
www.ksp.kit.edu

KIT – Universität des Landes Baden-Württemberg und
nationales Forschungszentrum in der Helmholtz-Gemeinschaft

Diese Veröffentlichung ist im Internet unter folgender Creative Commons-Lizenz
publiziert: http://creativecommons.org/licenses/by-nc-nd/3.0/de/

KIT Scientific Publishing 2013
Print on Demand

ISSN 1864-5933
ISBN 978-3-7315-0045-2

Multiscale Modeling of Cardiac Electrophysiology: Adaptation to Atrial and Ventricular Rhythm Disorders and Pharmacological Treatment

Zur Erlangung des akademischen Grades eines

DOKTOR-INGENIEURS

von der Fakultät für

Elektrotechnik und Informationstechnik

des Karlsruher Instituts für Technologie (KIT)

genehmigte

DISSERTATION

von

Dipl.-Ing. Mathias Wilhelms

geb. in Simmern/Hunsrück

Tag der mündlichen Prüfung: 6. Juni 2013
Referent: Prof. Dr. rer. nat. Olaf Dössel
Korreferenten: PD Dr. med. Eberhard P. Scholz
Prof. Dr. rer. nat. Uli Lemmer

Danksagung

Diese Arbeit ist während meiner Tätigkeit als wissenschaftlicher Mitarbeiter am Institut für Biomedizinische Technik des Karlsruher Institut für Technologie (KIT) zwischen Oktober 2009 und Juni 2013 entstanden. Viele Menschen, bei denen ich mich im Folgenden herzlich bedanken möchte, haben mich bei dieser Arbeit unterstützt.

An erster Stelle möchte ich Herrn Prof. Dr. rer. nat. Olaf Dössel für die kompetente Betreuung und die Übernahme des Hauptreferats danken. Seine hilfreichen Anregungen, sein großes Interesse an dieser Arbeit und die gemeinsamen fachlichen Diskussionen habe ich sehr geschätzt.
Weiterhin bedanke ich mich herzlich bei Herrn Prof. Dr. rer. nat. Uli Lemmer für die Übernahme des Korreferats und sein Interesse an der vorliegenden Dissertation.
Herzlicher Dank gilt auch Herrn PD Dr. med. Eberhard P. Scholz, der ebenfalls das Korreferat für diese Arbeit übernommen hat. Zusätzlich möchte ich mich für die exzellente und unkomplizierte Zusammenarbeit und die Bereitstellung medizinischer Messdaten bedanken.
Herrn Dr.-Ing. Gunnar Seemann danke ich für die hervorragende Betreuung in seiner Arbeitsgruppe Herzmodellierung, die hilfreichen Diskussionen und Anmerkungen, sowie für das Korrekturlesen dieser Arbeit.
Allgemein möchte ich allen Kollegen und auch ehemaligen Kollegen danken, die mir bei meiner Dissertation geholfen haben und diese auch Korrektur gelesen haben. Außerdem danke ich allen Studenten, die zu dieser Arbeit beigetragen haben, insbesondere Hanne Hettmann, sowie Axel Loewe und Jochen Schmid als Studenten und mittlerweile als Kollegen. Herrn Dr. rer.

nat. Matthias J. Krause danke ich für die umfassende Hilfe bei mathematischen Fragestellungen.

Furthermore, I would like to thank Dr. Molly Maleckar and Dr. Jussi Koivumäki for the great collaboration and the extremely nice visit to Oslo.

Besonderer Dank geht an meine Familie und meine Freunde für die liebevolle Unterstützung und Motivation während des Studiums und der Dissertation. Meinem Bruder danke ich ebenfalls für das Korrekturlesen dieser Arbeit.

Für einfach alles danke ich meiner Frau Nadine.

Teile dieser Arbeit wurden durch die Deutsche Forschungsgemeinschaft gefördert (DFG SE 1758/3-1).

Contents

Part III Results

8 Results: Parameter Adaptation of Electrophysiological Models   77

9 Results: Modeling Atrial Fibrillation 93

List of Abbreviations

ADP	Adenosine Diphosphate
AF	Atrial Fibrillation
AP	Action Potential
APD	Action Potential Duration
ATP	Adenosine Triphosphate
BCL	Basic Cycle Length
BSPM	Body Surface Potential Map
BZ	Border Zone
cAF	chronic Atrial Fibrillation
CHO	Chinese Hamster Ovary
CIZ	Central Ischemic Zone
CV	Conduction Velocity
DI	Diastolic Interval
DF	Dominant Frequency
ECG	Electrocardiogram
ERP	Effective Refractory Period
GA	Genetic Algorithm
hERG	human Ether-a-go-go-Related Gene
	(also termed KCNH2, which encodes I_{Kr})
KCNA5	*gene encoding I_{Kur}*

LAD	Left Anterior Descending (Coronary Artery)
LCx	Left Circumflex (Coronary Artery)
NCX	Sodium Calcium Exchanger
NKA	Sodium Potassium ATPase
NZ	Normal Zone
ODE	Ordinary Differential Equation
PMCA	Plasma Membrane Calcium ATPase
PSO	Particle Swarm Optimization
RCA	Right Coronary Artery
SR	Sarcoplasmic Reticulum
SS	Subspace
TMV	Transmembrane Voltage
TRR	Trust-Region-Reflective
VW	Vulnerable Window
WL	Wave Length
WT	Wild Type
ZF	Zone Factor

Cardiac Ion Currents:

I_{bCa}, I_{bCl}, I_{bNa}	Background Calcium/Chloride/Sodium Current
I_{CaL}	L-Type Calcium Current
$I_{Cl(Ca)}$	Calcium-Activated Chloride Current
I_{di}	Calcium Diffusion Current
I_f	Hyperpolarization-Activated Inward Potassium Current
$I_{K,ATP}$	ATP Sensitive Potassium Current
I_{K1}	Inward Rectifier Potassium Current
I_{KACh}	Acetylcholine-Acitvated Potassium Current
I_{Kp}	Plateau Potassium Current

I_{Kur}, I_{Kr}, I_{Ks} Ultra-Rapid/Rapid/Slow Delayed Rectifier Potassium Current

I_{leak} Calcium Leak Current

I_{Na} Fast Sodium Current

I_{NCX} Sodium Calcium Exchange Current

I_{NKA} Sodium Potassium ATPase Current

I_{PMCA} Plasma Membrane Calcium ATPase Current

I_{rel} Calcium Release Current

I_{to} Transient Outward Potassium Current

I_{tr} Calcium Transfer Current

I_{up} Calcium Uptake Current

1

Introduction

1.1 Motivation

In general, the synchronized contraction of the heart supplying the entire body with blood results from underlying processes of cardiac electrophysiology. Two pathologies investigated in this thesis, i.e. atrial fibrillation (AF) and acute cardiac ischemia, impair the functionality of the heart. AF is the most common arrhythmia causing uncoordinated activation of the atria with high beating frequencies. This atrial rhythm disorder slowly progresses over time, i.e. AF finally becomes chronic. Due to this, the risk for thromboembolism is significantly increased leading e.g. to strokes. In case of acute cardiac ischemia, the blood supply of the myocardium is reduced. Therefore, it is the most common cause of death in western industrialized countries. The effects of ischemia exacerbate over a short period in the range of few hours often leading to ventricular arrhythmias or lethal pump failure. Therefore, an early diagnosis and effective treatment of both pathologies is essential.

However, the complex underlying mechanisms responsible for the initiation, maintenance and also the termination of AF are not completely understood. Furthermore, the diagnosis of acute cardiac ischemia is based on shifts of the ST segment in the body surface electrocardiogram (ECG). The origin of these shifts, which can not be observed in more than one fourth of the patients with acute cardiac ischemia [1], has also not been investigated in its entirety yet. The impact of pharmacological therapy on human cardiac electrophysiology in general and additionally on the initiation or termination of arrhythmias is in the focus of current research as well.

These effects of pharmacological agents and atrial and ventricular rhythm disorders on cardiac electrophysiology can hardly be measured in human hearts, especially non-invasively. Therefore, reproducible *in silico* multiscale simulations ranging from the ion channel level up to the body surface ECG allow for testing a multitude of hypotheses without affecting the health of the patients. Different electrophysiological effects can be integrated into the models at the ion channel or single-cell level based on experimental data. Then, the complex anti- or proarrhythmic mechanisms can be investigated in detail at different simulation scales using quantitative measures.

In this way, the understanding of cardiac arrhythmias and the early diagnosis of these pathologies can be improved. Furthermore, also the development of pharmacological agents can be supported by the findings of these simulations reducing the risk and costs of clinical trials. In the future, these multiscale simulations might enable patient-specific therapy with tailored pharmacological agents for e.g. patients with mutations of ion channels causing AF.

1.2 Focus of the Thesis

Several aspects at different levels of multiscale simulations ranging from the ion channel up to the torso are in the focus of this thesis.

At the ion channel level:

- Implementation of a framework for the parameter adaptation of models of cardiac ion channels to voltage or patch clamp measurement data based on a combination of derivative-free and a trust-region-reflective optimization algorithm
- Integration of voltage and patch clamp measurement data of wild type (WT) and mutated channels known to be associated with AF into models of I_{Kr} and I_{Kur}

At the single-cell level:

- Adaptation of models of human atrial electrophysiology to effects of chronic AF (cAF)
- Modification of a model of human ventricular electrophysiology to effects of acute cardiac ischemia
- Integration of influence of three pharmacological agents on the control and pathologic models of atrial and ventricular electrophysiology and investigation of the resulting changes of cardiac action potentials

At the tissue level:

- Investigation of the restitution properties of cardiac excitation propagation of the different modified models in 1D tissue strands
- Analysis of initiation and termination of reentrant circuits simulated in a 2D tissue patch using the different atrial models

At the organ level:

- Studying the influence of acute cardiac ischemia or pharmacological treatment on excitation propagation simulated in a 3D ventricular model.

At the torso level:

- Investigation of changes of the body surface ECGs due to the effects of acute cardiac ischemia or pharmacological therapy

1.3 Structure of the Thesis

Part I outlines medical and technical background information:

- **Chapter 2** gives a brief introduction to cardiac anatomy, electrophysiology and related measurement techniques. Furthermore, the pathological changes due to AF and acute cardiac ischemia are introduced. Then, the impact of pharmacological therapy on cardiac electrophysiology is delineated.
- **Chapter 3** describes the electrophysiological models of human cardiac myocytes used in this thesis and previous modifications of these models. Furthermore, mathematical descriptions of cardiac excitation propagation and the forward calculation of body surface potentials are outlined.

Part II presents the utilized methods:

- **Chapter 4** describes the parameter adaptation framework including different optimization algorithms. Parts of this work were created during the supervised bachelor thesis of Niko Konrad [2], the diploma thesis of Jochen Schmid [3] and the master thesis of Axel Loewe [4].
- **Chapter 5** presents five models of human atrial electrophysiology used for the simulation of AF and the different investigations to assess the arrhythmic potency of the models. Parts of this work were created in the course of the supervised bachelor thesis of Hanne Hettmann [5] and Vanessa Lupici-Baltzer [6].

- **Chapter 6** introduces the modifications of the ventricular model to describe different stages of acute cardiac ischemia and the creation of heterogeneous ischemic tissue. The supervised bachelor thesis of Axel Loewe [7] and the student research project of Franziska Grimm [8] partly contributed to this work.

- **Chapter 7** describes the changes of the atrial and ventricular models to simulate the effects of three pharmacological agents. Furthermore, methods for the assessment of the arrhythmic potency of the compounds are shown. Parts of this work were created during the supervised student research project of Stefan Ponto [9] and Lukas Holl [10].

Part III

- **Chapter 8** presents the results of the parameter adaption using different optimization algorithms. Parts of these results were shown at a scientific conference [11] and are submitted to a scientific journal [12].

- **Chapter 9** shows the benchmark of different models of human atrial electrophysiology, especially regarding their ability to reproduce the effects of cAF, which was published in parts in a scientific journal [13]. Furthermore, the impact of ion channel mutations on the initiation of AF was investigated.

- **Chapter 10** provides the results of simulating the impact of different stages of acute cardiac ischemia, which was presented in parts on scientific conferences [14–16] and in a scientific journal [17].

- **Chapter 11** presents the results of simulations of the effects of three pharmacological agents on atrial and ventricular electrophysiology. Parts of this work were published in a scientific journal [18].

Chapter 12 summarizes the main findings presented in this thesis and gives an outlook on future work in this field.

Fundamentals

2

Medical Background

This chapter presents the basic medical background information useful for the understanding of the following technical chapters. A brief introduction to cardiac anatomy, electrophysiology and measurement techniques of the electrical activity of the heart is given. Then, pathological changes due to AF and acute cardiac ischemia are described. Finally, the impact of pharmacological therapy on cardiac electrophysiology is outlined. Furthermore, additional literature on these topics is referenced in the corresponding sections.

2.1 Cardiac Anatomy

The human heart, which has about the size of a closed fist, is the most important part of the circulatory system. It is responsible for pumping oxygen-rich and deoxygenated blood to supply all organs with nutrients and O_2, as well as to remove CO_2 and waste products, respectively. For this purpose, the heart (compare Fig. 2.1) is divided into two halves, which are separated by the septum. The left atrium receives the oxygen-rich blood from the lungs through the pulmonary veins and fills the left ventricle, which in turn pumps the blood through the aorta into the systemic circulation. The blood flow is regulated by the mitral valve between the left atrium and ventricle and the aortic valve between the left ventricle and the aorta. The left ventricular wall is much thicker than the right ventricular wall due to the higher pressure in the systemic circulation compared to the pulmonary circulation. The right atrium receives the deoxygenated blood through the superior and inferior vena cava and fills the right ventricle, which pumps the blood through the pulmonary artery to the lungs. The blood flow between the right atrium and

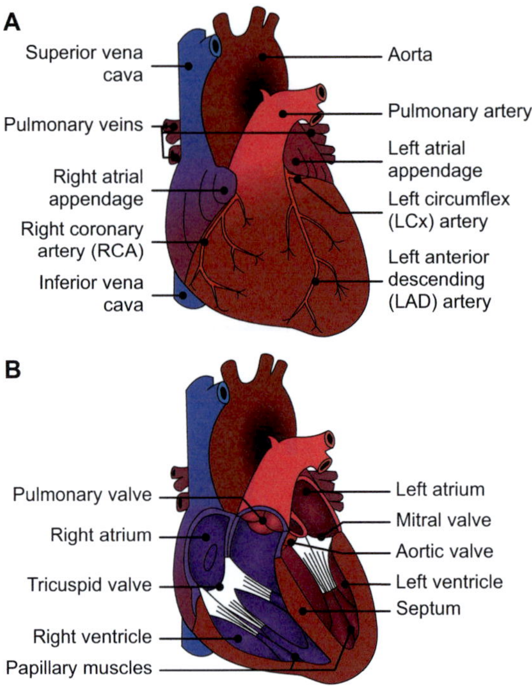

Figure 2.1. Schematic drawings of the anatomy of the human heart presenting the coronary arteries (A) and the interior (B).

ventricle is regulated by the tricuspid valve and between the right ventricle and the pulmonary artery by the pulmonary valve [19].

The heart itself is also supplied with oxygen and nutrients by the coronary arteries. In Fig. 2.1A, the main coronary arteries descending directly from the aorta are shown. The left coronary artery, which passes the left atrial appendage, branches into the left anterior descending (LAD) artery and the left circumflex (LCx) artery. The anterior part of both ventricles including the interventricular septum is supplied by the LAD, whereas the LCx provides the left atrium and ventricle with oxygen-rich blood. The right coronary artery (RCA) passes the right atrial appendage and supplies the right atrium and ventricle, as well as the posterior part of both ventricles including the interventricular septum [20].

Cardiac myocytes, which present a roughly cylindrical shape, are arranged as fibers in the atrial and ventricular walls. Partly due to this fiber orientation, the conduction velocity (CV) of cardiac excitation propagation is anisotropic being faster in direction of the fiber than in transverse direction [21]. In the ventricles, the fiber orientation can be described by the helix angle α_1, which defines the inclination of the fibers from apex to base, and the transverse angle α_3, which quantifies the imbrication of the fibers between the endocardium and epicardium. According to Streeter [22], α_1 increases almost linearly from $-75°$ at the epicardium to $+55°$ across the ventricular wall. Furthermore, α_3 slightly varies from $-3°$ at the apex to $+3°$ at the base. A detailed description of the ventricular fiber orientation is given in [23].

2.2 General Cardiac Electrophysiology

This section summarizes the electrophysiological basics including the initiation of electrical activity of single cardiac myocytes and the propagation of cardiac excitation waves. Furthermore, the techniques measuring the electrical activity at the ion channel or single-cell level and at the body surface are outlined.

2.2.1 Electrophysiological Basics

The intracellular space of cardiac myocytes is separated from the extracellular space by a phospholipid bilayer, the cell membrane. As a consequence, different ion concentrations in the intra- and extracellular spaces can be maintained. This imbalance causes the transmembrane voltage (TMV), which is the difference between intra- and extracellular potential. Furthermore, the membrane contains embedded proteins, which serve as selectively permeable pathways for ions and other substances, for which the membrane is impermeable otherwise. These channels consist of cylindrical pore-forming α-subunits, which determine the diameter and selectivity of the channel. In addition, some ion channels have auxiliary subunits, called β, γ, etc., which can modify the function of the channel. Cardiac ion channels can switch between different states modifying their conductivity by changes in the conformation of the proteins, which is called gating. The channels can either be in a closed state blocking all currents or in an open state allowing

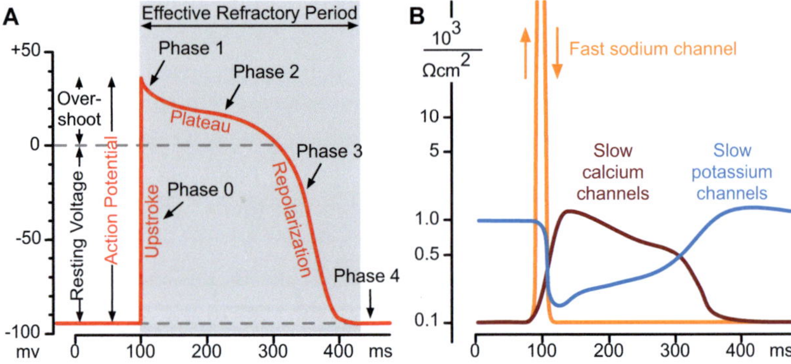

Figure 2.2. Phases of cardiac action potential (A) and the underlying transmembrane currents (B). Modified based on [26].

the passage of specific ions. Additionally, some channels exhibit a nonconducting inactivated state, which is entered after the activation. Most cardiac ion channels are voltage-gated, i.e. they change their state depending on the transmembrane voltage. The binding of specific substances to receptors governs the gating in case of ligand-gated ion channels. An overview of cardiac ion channels and the gating processes is given in [24]. In addition to ion channels, which allow passive ion transport along an electrochemical gradient, exchangers and ion pumps permit active transport against an electrochemical gradient. Energy is used for this process, e.g. by dephosphorylation of adenosine triphosphate (ATP) to adenosine diphosphate (ADP) [25].

Due to the imbalance between the intra- and extracellular ion concentrations in the resting state, the ions move passively from regions with higher concentration to lower concentration. Since the ions are charged particles, the electrical gradient increases with decreasing gradient. If both forces balance each other, the equilibrium Nernst voltage can be determined for the ion type X:

$$E_x = -\frac{RT}{z_x F} ln \left(\frac{[X]_i}{[X]_o} \right), \qquad (2.1)$$

with the gas constant R, the temperature T, the valence of the ion z_x, Faraday's constant F and the intra- and extracellular concentrations $[X]_i$ and $[X]_o$, respectively. In the resting state, mainly potassium channels are conductive (compare Fig. 2.2B) causing an outward current, so that the transmembrane voltage V_m is close to the potassium Nernst voltage. If an external stimulus depolarizes the membrane above a threshold voltage, an action potential (AP) is initiated and a fast sodium inward current further depolarizes the membrane. As a consequence, a rapid upstroke of the transmembrane voltage with an overshoot to positive values can be observed (Fig. 2.2A, phase 0). The sodium channels quickly inactivate and remain in this state until the transmembrane voltage falls below another threshold value. The interval, during which no further AP can be propagated to neighboring cells, is called effective refractory period (ERP). Inactive sodium channels prevent the initiation of further APs in this case. Conductive potassium channels and a decrease of the sodium channel conductivity cause a slight repolarization (phase 1). In the meantime, the potassium conductivity is reduced and calcium channels open. The resulting calcium inflow triggers an additional calcium release from the sarcoplasmic reticulum (SR), which is a special intracellular calcium storage. This leads to the plateau phase of the AP (phase 2). The slow calcium channels inactivate and the potassium conductivity increases again, so that the membrane is repolarized (phase 3). Finally, the resting membrane voltage is reached (phase 4) and the imbalance of ion concentrations is restored by exchangers and ion pumps, as e.g. the sodium calcium exchanger (*NCX*), or the sodium potassium ATPase (*NKA*) [25]. The interval between the upstrokes of two consecutive APs is called basic cycle length (BCL). The time between the end of a previous AP and the upstroke of the following AP, which is called diastolic interval (DI), is calculated as the difference between the BCL and the action potential duration (APD).

Adjacent cardiac myocytes are connected to each other by a different type of integral membrane proteins, so-called gap junctions. Connexons, which consist of six connexins, form tubular structures between the intra- and extracellular space. The connexons of adjacent cells face each other bridging the gap between them, so that the intracellular spaces of the myocytes are connected. Different types of connexins with varying size of the tubes exist. Therefore, certain types of signaling molecules or ions can be transferred between the cells. In this way, the electrical excitation propagates from cell to cell until the entire heart, which is a so-called syncytium, is activated. The density and type of the gap junctions determines the CV in the tissue and is partly responsible for the anisotropy of the CV.

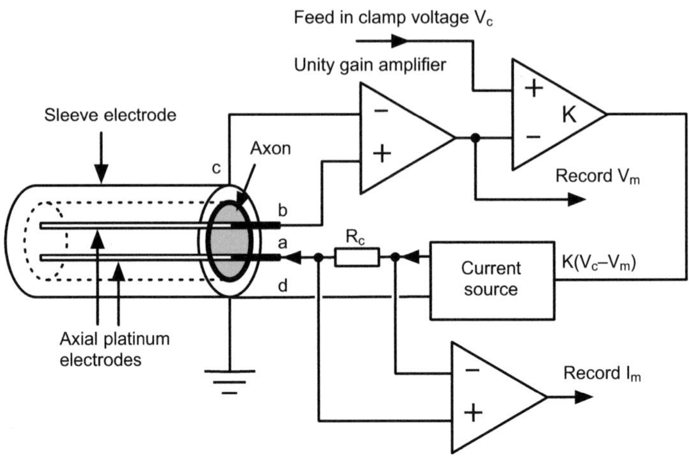

Figure 2.3. Schematic voltage clamp measurement circuit. Difference between transmembrane voltage V_m, which is measured using electrodes (b) and (c), and the clamp voltage V_c is amplified and a current source feeds the current to electrodes (a) and (d). The resulting current transmembrane current I_m is measured. Adopted from [27].

2.2.2 Voltage Clamp and Patch Clamp Technique

In 1949, Cole [28] and Marmont [29] developed the voltage clamp technique, which allows the measurement of ion currents across the cell membrane. Three years later, Hodgkin and Huxley [30] used this technique to quantitatively analyze the underlying ion currents, on which the action potentials of giant squid axons are based. The principle behind the voltage clamp technique is sketched in Fig. 2.3. One platinum electrode injected into the intracellular space of the cell (b) and a reference electrode in the extracellular space (c), which is a bath surrounding the cell, are used to measure the transmembrane voltage V_m. The difference between the defined clamp voltage V_c and V_m is amplified afterwards. A voltage-controlled current source then supplies the electrodes (a) and (d) resulting in the transmembrane current I_m, which is measured using a calibrated resistance R_c. In this way, the negative feedback loop minimizes the difference between the transmembrane voltage and the impressed clamp voltage until they are equal. Since the clamp voltage is constant except for the instant changes at the voltage steps, the membrane capacity can be neglected, so that only the ionic currents across the membrane are measured. Due to this, the response of all cardiac ion channels of a myocyte to certain transmembrane voltages can be determined. In order to record only one specific ion current, all other currents have to be blocked using pharmacological agents, which might also influence the current of interest. Another option is to use cells, which do not express ion channels on their own. These cells are transfected with genes encoding the ion channel of interest, which is expressed afterwards.

Neher and Sakman [31] developed the patch clamp technique in 1976, which allows to measure ion currents of few or even a single ion channel. Instead of two electrodes injected into the cell, a glass micropipette is attached to the cell surface, so that a small patch of the cell is sucked into the pipette. Then, different recording methods exist allowing to measure e.g. only one ion channel at this patch, or the membrane of the whole cell. In case of the whole-cell recording, the membrane patch is ruptured, so that a direct connection between the inside of the pipette and the intracellular space of the cell is created. Similar to the voltage clamp technique, a feedback current flowing through the pipette is used to compensate the difference between the defined clamp voltage and V_m. However, the series resistance, which is the resistance between the membrane and the input of the amplifier, has to be compensated, as described in detail in e.g. [32]. An overview of the voltage clamp and patch clamp technique including the different recordings is given in [27]. The voltage clamp and whole-cell patch clamp recordings shown in this thesis were provided by the University Hospital Heidelberg, Department of Internal Medicine III.

2.2.3 Cardiac Conduction System and Genesis of the Electrocardiogram

As described in section 2.2.1, cardiac myocytes are connected by gap junctions allowing propagation of electrical excitation waves. These signals are reflected in the electrical potential at the body surface, which is measured by the ECG. A narrow depolarization front causes an intracellular current in direction of the propagation from the already activated regions to the not yet activated regions. In the meantime, a sodium influx from the extracellular space can be observed at the back of the front, whereas a capacitive outflow at the front further depolarizes the neighboring cells. As a conse-

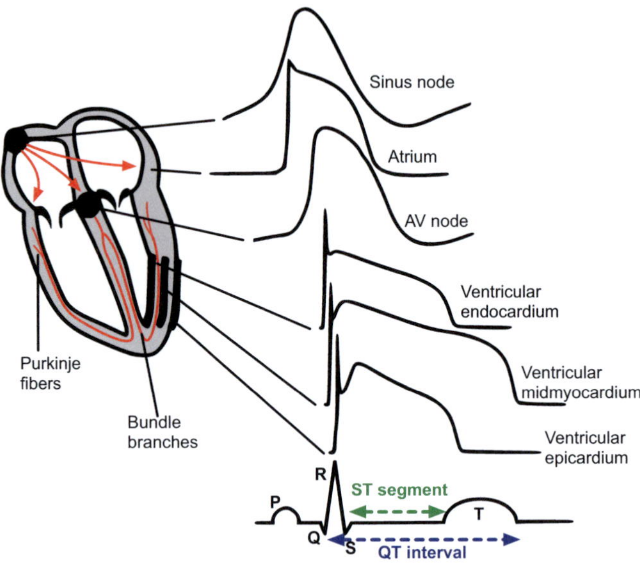

Figure 2.4. Schematic conduction system and cardiac APs from different regions of the heart resulting in the ECG at the body surface. Modified based on [33].

quence, the resulting double layer source has a positive polarity in direction of the propagation of the front and consequently negative polarity in the opposite direction. The repolarization wave, which is much broader than the depolarization front, causes an intracellular current from the still active to the repolarized regions and a potassium outflow at the wave back. A capacitive inflow as local circuit current is present at the border to still active tissue. Consequently, a repolarization wave leads to a double layer source with negative polarity in direction of the wave and positive polarity in the opposite direction. The effects responsible for the genesis of electrical signals at the body surface are described in detail in [34].

In the human heart, various cell types exist in different regions of the heart shown in Fig. 2.4, which cause the electrical excitation of the entire organ. During normal sinus rhythm, the electrical activity is initiated at the sinus node. There, specialized pacemaker cells without stable resting membrane voltage automatically trigger APs. The excitation is propagated across the entire atria resulting in the P wave. The atria are electrically isolated from the ventricles except for the atrioventricular (AV) node. After a short delay at the AV node, the excitation wave propagates along the bundle of His and the bundle branches towards the Purkinje fibers, which are specialized pathways with high CV. In this way, rapid depolarization of the entire ventricles occurs, which reflects in the QRS complex of the ECG. The more or less homogeneous plateau phase in the ventricles leads to the ST segment of the ECG, which is nearly a zero baseline. The heterogeneous repolarization in the ventricles causes the subsequent T wave [20]. In the clinical practice, a 12-lead system is commonly used to derive an ECG. It consists of three electrodes, which are used as unipolar (Goldberger aV_R, aV_L and aV_F) and bipolar (Einthoven I to III) leads, close to or on both arms and the left leg, as well as six unipolar chest leads (Wilson V_1 to V_6) close to the heart [34].

2.3 Atrial Fibrillation

AF is a rhythm disorder affecting more than 1% of the population [35]. Since the prevalence increases with increasing age, even more people will suffer from this pathology in the future as the population is getting older [36]. This supraventricular tachyarrhythmia presents an uncoordinated activation of the atria with very high beating frequencies between 400 to 600 bpm [37]. AF is responsible for up to one fourth of all strokes [38]. Additionally, it is generally associated with hypertension, diabetes, heart failure and other cardiovascular diseases and furthermore increases the mortality [39]. However, the underlying mechanisms responsible for AF are not completely understood yet [40]. The focus of pharmacological and interventional therapy of AF is on the management of heart rate and rhythm, as well as on the reduction of the thromboembolic risk. Admittedly, current pharmacological treatment is still not efficient enough and has many side effects [41]. As a consequence, the management of AF has to be improved starting earlier and being more comprehensive [35]. Detailed information on AF is given in [42]. However, only the aspects of AF investigated in this thesis, i.e. remodeling due to cAF and genetic defects favoring AF, are described in the following.

2.3.1 Remodeling due to Chronic AF

AF is a cardiac arrhythmia, which slowly progresses over time. It starts with few paroxysmal episodes and finally becomes permanent/chronic [43]. AF itself initiates mechanisms, which promote the occurrence and maintenance of AF. This principle that "AF begets AF" was established by Wijffels et al. [44] in 1995. Due to the rapid rates during AF, the affected tissue undergoes electrical remodeling, which influences several ion channels and gap junctions. This remodeling causes a shortening of the APD and ERP, as well as a loss of the rate adaptation of the ERP [42]. Electrical remodeling due to cAF alters the expression or phosphorylation of several cardiac ion channels. The L-type calcium current I_{CaL} and the transient outward potassium current I_{to} are both reduced by approximately 65% [45, 46]. Furthermore, cAF-induced remodeling decreases the ultra-rapid delayed rectifier potassium current I_{Kur} by 49% [46]. In contrast, the inward rectifier potassium current I_{K1} is increased by around 110% [47]. Changes of further ion currents, which were neglected in this work, as e.g. the acetylcholine-activated potassium channel I_{KACh}, are summarized in [42].

In addition to ion channels, also gap junctions may be influenced by cAF. However, the findings on gap junctional remodeling reported in literature are inconsistent [48] showing decrease or even increase of connexin expression and conflicting spatial distributions. However, a significant decrease by 53% of the expression of connexin 40 (Cx40), which can mainly be found in the atria, is observed in e.g. [49].

2.3.2 Genetic Defects Favoring AF

In general, AF is associated with structural heart diseases, such as valvular heart disease, coronary artery disease, heart failure, hypertension, and so on. However, AF is sometimes diagnosed in patients, especially if they are young, who do not present these typical cardiovascular comorbidities. In this case, mutations of genes encoding proteins of the cardiac cell membranes are associated with AF instead [37]. This type of AF can be classified as either familial AF, if close relatives are known to have the same form of AF, or as sporadic AF, if no relatives are affected [50]. Mutations of genes encoding cardiac ion channels, connexins, atrial natriuretic peptide, ryanodine receptors and other proteins are associated with AF [37]. An overview of selected genetic defects of ion channels and connexins associated with AF is given in Table 2.1. Different genes encoding cardiac ion channels are affected by either gain-of-function or loss-of-function mutations increasing or decreasing the resulting currents. Mutations affecting the rapid delayed rectifier potassium current I_{Kr}, the slow delayed rectifier potassium current I_{Ks}, I_{Kur}, I_{K1}, the fast sodium current I_{Na}, and I_{CaL} are reported [37, 51]. In Fig. 2.5, an exemplary mutation of KCNH2, which is also termed human ether-a-go-go-related gene (hERG), is shown. In this case, the channel consists of six transmembrane segments S1 to S6. At codon 588, the mutation N588K, which is associated with AF and the short QT syndrome, changes the amino acid asparagine (N) to lysine (K), explaining the nomenclature of the mutations. Although numerous genetic defects associated with AF have been identified already in recent years, further research is needed to investigate the mechanisms responsible for the initiation and persistence of AF [42].

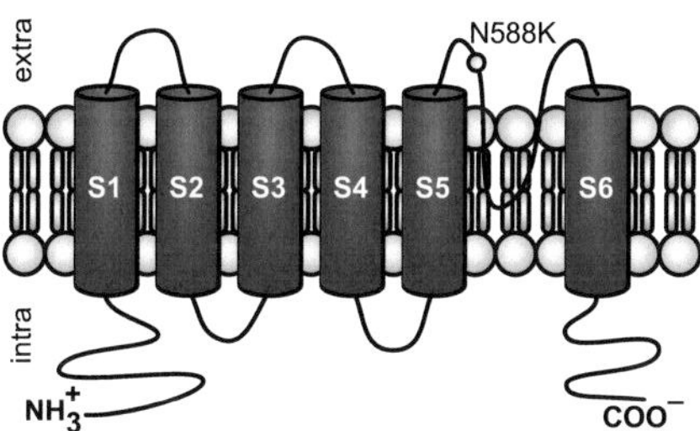

Figure 2.5. Schematic visualization of the hERG channel consisting of six transmembrane segments. Mutation N588K causes a substitution of asparagine (N) by lysine (K) at codon 588 between segments S5 and S6.

Table 2.1. Selected genetic defects associated with AF. Modified based on [37, 51].

type	channel	gene	chromosome	variants	effects
potassium	I_{Kr}	KCNE2	21q22	R27C	gain-of-function
channel	I_{Kr}	KCNH2[1]	7q35-q36	N588K	gain-of-function
	I_{Ks}	KCNQ1	11p15	S140G, V141M	gain-of-function
	I_{Ks}	KCNE5	Xq22.3	L65F	gain-of-function
	I_{Ks}	KCNE3	11q13-q14	R53H	unknown
	I_{Kur}	KCNA5	12p13	E375X, T527M, A576V, E610K	loss-of-function
	I_{K1}	KCNJ2	17q23-q24	V93I	gain-of-function
sodium	I_{Na}	SCN5A	3p21	M1875T	gain-of-function
channel	I_{Na}	SCN5A	3p21	D1275N, N1986K	loss-of-function
	I_{Na}	SCN1B	19q13.1	R85H, D153N	loss-of-function
	I_{Na}	SCN2B	11q23	R28Q, R28W	loss-of-function
calcium	I_{CaL}	CACNA1C	12p13.1	A39V, G490R	loss-of-function
channel	I_{CaL}	CACNB2	12p13.1	S481L	loss-of-function
gap junction	Cx40	GJA5	1q21.1	P88S, M163V, G38D, A96S	negative effect on gap junctions
	Cx43	GJA1	6q22-q23	c.932delC	negative effect on gap junctions

[1] also known as human ether-a-go-go-related gene (hERG).

2.4 Acute Cardiac Ischemia

Acute cardiac ischemia is caused by the partial or complete occlusion of a
coronary artery, which normally supplies the heart with oxygen and nutri-

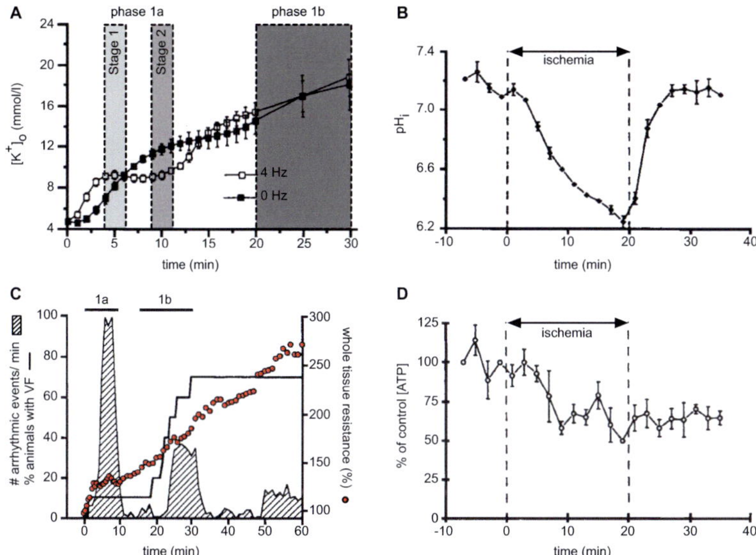

Figure 2.6. Main effects of acute cardiac ischemia. Effects of hyperkalemia measured in rat hearts [54] (A), acidosis observed in ferret hearts [55] (B), cell-to-cell uncoupling and occurrence of arrhythmias investigated in swine [56] (C), lack of ATP caused by hypoxia observed in ferret hearts [55] (D) at different stages of ischemia.

ents. As a consequence of this shortage, different spatio-temporal changes influence the electrophysiology, which can even be observed in the body surface ECG. Only the underlying effects having a direct impact on cardiac electrophysiology are described in the following, whereas a general overview and further detailed information on e.g. metabolic effects of acute cardiac ischemia can be found in [52, 53].

2.4.1 Electrophysiological Effects of Acute Cardiac Ischemia

During the first thirty minutes of ischemia, two distinct phases, during which arrhythmias occur most often, can be identified. These phases, which can be seen in Fig. 2.6C, are called phase 1a between around $2 - 10\,\text{min}$ and phase 1b between $20 - 30\,\text{min}$ after the onset of ischemia [53]. The electrophysiological effects of phase 1a can be subdivided into stage 1 at around $5\,\text{min}$ and stage 2 at around $10\,\text{min}$ after the onset of ischemia effects. Generally, acute cardiac ischemia elevates the resting membrane voltage $V_{m,rest}$, decreases the AP amplitude, shortens the APD, but prolongs the ERP and reduces the CV [52]. The main underlying effects of the first thirty minutes of ischemia are described in the following.

An accumulation of extracellular potassium is observed after the onset of ischemia, which is shown in Fig. 2.6A. In general, this extracellular hyperkalemia can be explained by activation of an ATP sensitive potassium current $I_{K,ATP}$ due to loss of ATP [57], as well as by altered sodium currents, which cause an increase of the intracellular sodium concentration $[Na^+]_i$. This increased $[Na^+]_i$ results in passive potassium efflux preserving electroneutrality [58]. After a rapid increase of the extracellular potassium concentration $[K^+]_o$, a short plateau phase can be observed, during which the concentration remains nearly constant. This can be explained by temporary reactivation of NKA by glycolysis [59].

After the onset of ischemia, the pH falls towards more acidic values, which can be seen in Fig. 2.6B. This acidosis can be explained by an increased production of protons due to anaerobic glycolysis. Furthermore, less protons are removed from the ischemic regions, due to an insufficient removal of CO_2. As a consequence, several ion channels and gap junctions are inhibited by the decreased pH, since these proteins behave as enzymes [53].

An occlusion of a coronary artery, which results in reduced availability of O_2 (hypoxia), also causes a lack of ATP (compare Fig. 2.6D), since anaerobic phosphorylation and local energy storages cannot supply the demand for ATP. This reduction in ATP on the one hand increases $I_{K,ATP}$, which is inhibited by physiological concentrations of ATP [60], and additionally inhibits the active transport mechanisms of the membrane, as e.g. NKA [53].

In Fig. 2.6C, the increase of the whole tissue resistance due to acute cardiac ischemia is shown. Several factors reducing the gap junction conductivity, as e.g. increased $[Na^+]_i$ and $[Ca^{2+}]_i$, as well as acidosis, have been identified [61]. This cell-to-cell uncoupling, which starts to increase after $20-30$ min of ischemia, is associated with phase 1b arrhythmias [62].

2.4.2 Heterogeneous Ischemic Tissue

Effects of acute cardiac ischemia generally do not occur in the entire heart, which is called global ischemia, but only regionally depending on the oc-

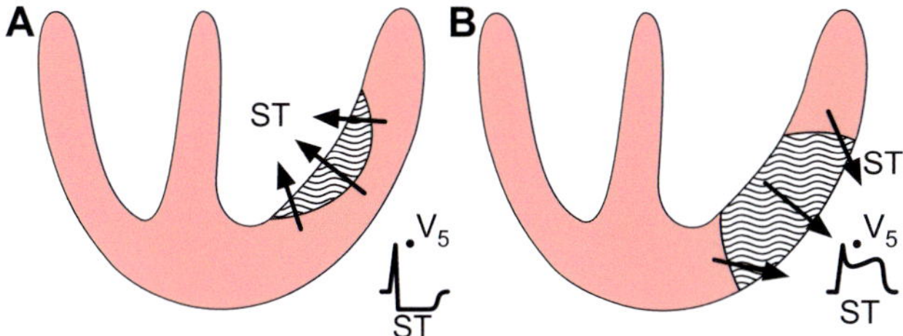

Figure 2.7. Effects of acute subendocardial and transmural ischemia on the ECG. Subendocardial ischemia (A) causes injury currents directed to the ischemic endocardium resulting in ST segment depression. In case of transmural ischemia, the currents are directed from the less injured endocardium to the epicardium causing ST segment elevation. Adopted from [69].

clusion site. Therefore, heterogeneous distribution of ischemia effects can be observed. A border zone (BZ) (<10 mm in width) is located between the healthy tissue and the central ischemic zone (CIZ), where the effects are most pronounced [63, 64]. The various ischemia effects develop differently across the BZ [65].

Additionally, the effects of acute cardiac ischemia vary also transmurally. Since the distance to the coronary arteries is greatest in the endocardium and the blood flow, contraction and therefore also the metabolic activity is higher in the endocardium of a healthy heart [66], ischemia effects occur in this inner layer first. The effects of this subendocardial ischemia spread towards the epicardium becoming transmural, if the occlusion worsens [67]. Nevertheless, ischemia effects are more pronounced in the epicardial tissue due to a higher sensitivity of $I_{K,ATP}$ to loss of ATP there [68].

2.4.3 Effects of Ischemia on the ECG

The electrophysiological changes due to acute cardiac ischemia are reflected in the body surface ECG changing e.g. the QRS complex or the T wave [70, 71]. However, the most pronounced effects, i.e. shifts of the ST segment of the ECG, are used as a diagnostic marker of ischemia and myocardial infarction [72]. The effects, which are assumed to be responsible for these shifts are presented in Fig. 2.7. Injury currents flow from the healthy epicardial towards the ischemic endocardial tissue in case of subendocardial ischemia, causing ST segment depression in leads close to the ischemic region. However, transmural ischemia results in injury currents directed from the less injured endocardial to the more affected epicardial tissue. Due to this, a pronounced ST segment elevation can be observed in leads close to the ischemic region in this case [73].

2.5 Pharmacological Effects on Cardiac Electrophysiology

The effects of pharmacological therapy on cardiac electrophysiology are outlined in this section. Initially, a short overview of pharmacodynamics, i.e. how the binding of pharmacological agents to receptors can be modeled, and the classification of antiarrhythmics depending on their impact on cardiac electrophysiology is given. Then, the three compounds amiodarone, dronedarone and cisapride, which were investigated in this thesis, are described.

2.5.1 General Mechanisms and Classification of Antiarrhythmic Agents

In general, the binding of a ligand L to a receptor R can be described by the law of mass action according to following equation:

$$L + R \underset{k_{off}}{\overset{k_{on}}{\rightleftharpoons}} LR, \tag{2.2}$$

where k_{on} is the association rate constant, with which the receptor-ligand complex LR is formed, and k_{off} is the dissociation rate constant of the opposite process [75]. In case of pharmacological agents, the binding and unbinding of the compound to receptors of ion channels is described by this equation. However, this simple model does not consider the effects of more

Table 2.2. Classification of antiarrhythmics according to Vaughan Williams [74].

class	action	effect
I	inhibition of sodium channel	decrease of upstroke (phase 0 of AP)
II	beta-blocking	decrease of heart rate
III	inhibition of outward potassium channel	prolongation of APD and refractory period
IV	inhibition of inward calcium channel	reduction of plateau phase (phase 2 of AP)

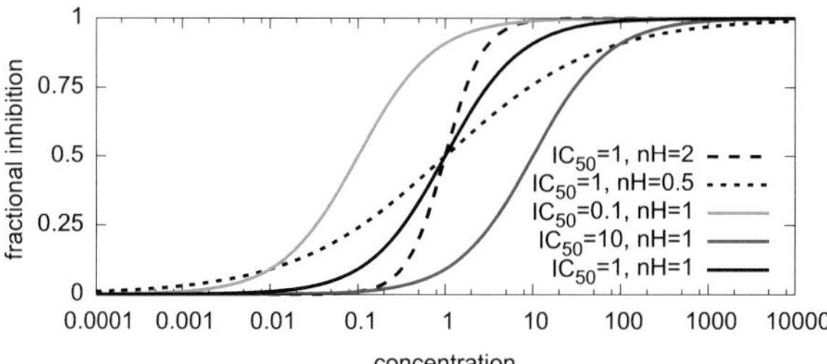

Figure 2.8. Hill curves with different IC_{50} and nH describing fractional inhibition of ion currents depending on the concentration of the pharmacological compound.

complex binding mechanisms, such as cooperativity, i.e. the binding affinity of a receptor is increased or decreased, if a ligand is already bound to a receptor. Hill [76] described this behavior in case of the binding of oxygen to hemoglobin in 1910. Accordingly, the equation describing the fractional inhibition of ion channels by a pharmacological agent L is named Hill equation. This fractional inhibition corresponds to the reduction of the maximal ion channel conductivity:

$$\text{fractional inhibition} = \frac{1}{1 + \left(\frac{IC_{50}}{[L]}\right)^{nH}}, \tag{2.3}$$

with IC_{50} being the half maximal inhibitory concentration, at which the resulting ion currents are inhibited by 50%, $[L]$ being the concentration of the pharmacological agent and nH being the Hill slope. $nH > 1$ indicates positive cooperativity, i.e. the affinity of the the receptor is increased upon binding of the ligand, whereas $nH < 1$ indicates negative cooperativity, i.e. the affinity of the receptor is reduced upon binding of the ligand [75]. In this way, the values of IC_{50} and nH describe the sigmoidal binding curve and therefore the dose-dependent inhibition of ion channels by pharmacological agents.

Antiarrhythmic agents can be classified according to their effect on cardiac electrophysiology. The different classes and corresponding effects are summarized in Table 2.2. The effects range from inhibition of cardiac ion channels, which can immediately be seen in the APs of the affected myocytes, to inhibition of the sympathetic nervous system. Sodium channel blocker

Table 2.3. Inhibition of cardiac ion currents due to the antiarrhythmic agent amiodarone.

current	IC$_{50}$ (μM)	nH	source
I_{Na}	1.4	1.0	[77]
I_{NKA}	15.6	1.0	[78]
I_{Kr}	0.8/0.07	1.3/1.1	[79]/[80]
I_{Ks}	3.84	0.63	[81]
I_{CaL}	5.8/0.55	1.0/0.26	[82]/[83]
I_{NCX}	3.3	1.0	[84]

are grouped in class I, potassium channel blocker in class III and calcium channel blocker in class IV. β-blocker inhibiting β-adrenergic receptors, which mainly influence the heart rate and blood pressure, are assigned to class II [74].

2.5.2 Antiarrhythmic Agent Amiodarone

Amiodarone is a class III antiarrhythmic agent, which is used for the management of ventricular tachycardia or ventricular fibrillation due to myocardial ischemia and infarction [85], as well as for the treatment of AF [86]. Although amiodarone is classified as class III antiarrhythmic agent due to its effective inhibition of potassium channels, it furthermore exhibits properties of the other classes, since it is a multichannel blocker. An overview of the affected ion currents is given in Table 2.3. The therapeutic concentrations of amiodarone range between 1 and 3 μM [87]. Therefore, effects on other channels with IC_{50} values far above the therapeutic concentrations, as e.g. I_{K1} with an IC_{50} of 92.14 μM [88], are neglected in this thesis. In general, the values describing the inhibition of the ion channels reported in literature vary significantly depending on the species and temperature of the cells that were used for the measurements. In case of I_{Kr} and I_{CaL}, these differences are most pronounced. Additionally, the effects of acute and chronic administration of amiodarone vary also [87]. As a consequence, either an increase of the APD, no change, or even a shortening is reported depending on the study and the species investigated [87]. Furthermore, no significant changes of the QT interval are visible after acute administration [89], whereas chronic treatment leads to a significant prolongation of the QT interval and the ERP [90, 91]. Nevertheless, the occurrence of torsade de pointes arrhythmias is rare [92]. However, corneal micro-deposits can be observed in most of the patients chronically treated with amiodarone [92]. Furthermore, amiodarone can induce severe thyroid dysfunction, which is the most common reason for discontinuation of chronic administration of amiodarone, due to its structural similarity to the thyroid hormone thyroxine [93]. Additional experimental data of the inhibition of cardiac ion channels are listed in [9, 94]. Further general information on amiodarone is given in e.g. [87, 95] and the different effects on atrial and ventricular electrophysiology, which are reflected in a prolongation of the APD predominantly in the atria, are described in [96].

2.5.3 Antiarrhythmic Agent Dronedarone

Dronedarone, which is a derivative of amiodarone, has been approved in the USA in 2009 [101]. The structure of amiodarone was systematically changed in order to reduce adverse side effects. On the one hand, iodine was removed to eliminate thyroid toxicity and on the other hand, the lipophilicity was decreased to reduce the accumulation in tissue [101]. Dronedarone is advised for the treatment of non-permanent AF, since recent studies showed that dronedarone is not as effective as amiodarone in the treatment of cAF [102]. Nevertheless, dronedarone, which is also a class III antiarrhythmic agent [103], presents similar effects on various cardiac ion channels as amiodarone [104]. The main cardiac ion channels inhibited by dronedarone are listed in Table 2.4. The steady state plasma concentrations are reported to be between 84 and 147 ng/ml [105], which corresponds to around 150 to 270 nM. Therefore, the impact on ion channels, which presented much higher IC_{50} values, as e.g. I_{K1} with an $IC_{50} > 30\,\mu$M [98], was neglected in the present work. Chronic treatment of rabbit atrial tissue with dronedarone increased the APD and ERP [106]. Additional experimental values describing the inhibition of cardiac ion channels are listed in [10]. Further general information on dronedarone is given in [100, 104].

Table 2.4. Inhibition of cardiac ion channels due to the antiarrhythmic agent dronedarone.

current	IC_{50} (μM)	nH	source
I_{Na}	0.54	2.03	[97]
I_{Kr}	0.0591	0.8	[80]
I_{Ks}	5.6	0.51	[98]
I_{Kur}	1.0	1.0	[99]
I_{CaL}	0.83	2.75	[100]

Table 2.5. Inhibition of cardiac ion channel due to the gastroprokinetic agent cisapride.

current	IC$_{50}$ (μM)	nH	source
I_{Kr}	0.023	0.8	[107]

2.5.4 Gastroprokinetic Agent Cisapride

Cisapride, which is a gastroprokinetic agent, was administered for the facilitation and restoration of gastrointestinal motility [108] until its withdrawal from market in the USA in 1999. In many patients, prolongation of the QT interval, torsades de pointes and ventricular fibrillation could be observed, which led to discontinuation despite the benefit of this pharmacological agent [109]. At the ion channel level, inhibition of I_{Kr} (Table 2.5) is visible at therapeutic concentrations. The peak plasma concentration is in the range of 150 to 300 nM [110]. However, more than 95% of cisapride is bound to plasma proteins. The effective concentration inhibiting cardiac ion channels is therefore much lower [107]. Inhibition of ion channels with IC_{50} values much higher than the effective concentration was neglected in the present work, e.g. I_{CaL} presented an IC_{50} of 46.9 μM [111]. Prolongation of the APD and the QT interval was also observed in different species, as e.g. guinea pig, rabbit and dog [112, 113]. More information on the electrophysiological effects of cisapride is given in e.g. [94, 107].

3

Computational Cardiac Modeling

The background information on modeling cardiac electrophysiology is outlined in this chapter. Mathematical descriptions used for multiscale simulations, which range from the cell membrane of single myocytes up to models of cardiac tissue, are introduced. In this thesis, five models of human atrial electrophysiology and one model of ventricular electrophysiology were used. Furthermore, the anatomical model of the ventricles used in this thesis and the forward calculation of body surface ECGs is summarized. Further reading is referenced in the corresponding sections.

3.1 Modeling Electrophysiology of Cardiac Myocytes

The mathematical descriptions of the membranes of cardiac myocytes are mostly based on the findings of Hodgkin and Huxley [30]. In 1952, they investigated the membranes of giant squid axons and developed an electrical circuit representing these membranes, which is shown in Fig. 3.1. The

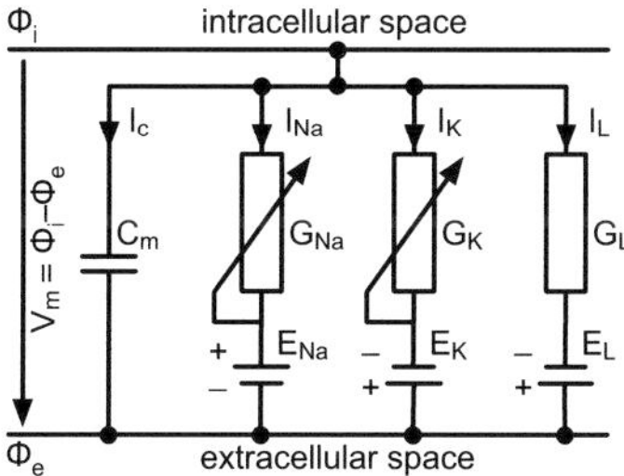

Figure 3.1. Electrical circuit representing the membrane of a giant squid axon according to Hodgkin and Huxley [30].

transmembrane voltage V_m, which is the difference between the intracellular potential Φ_i and the extracellular potential Φ_e, depends on the sum of ion currents I_{ion} and a capacitive current I_c, which results from the dielectric properties of the membrane. This can be described by the following ordinary differential equation (ODE):

$$\frac{dV_m}{dt} = -\frac{I_{ion} + I_{stim}}{C_m}, \tag{3.1}$$

where I_{stim} is a stimulus current and C_m the membrane capacity. This formulation is still used in current models of cardiac electrophysiology. Hodgkin and Huxley identified three currents contributing to I_{ion} in the giant squid axon: a sodium current I_{Na}, a potassium current I_K and a leakage current I_L [30]. The different ion currents I_x each depend on the maximal conductivity g_x, the product of gating variables γ_i describing the open probability of the channels, as well as the difference between the transmembrane voltage and the Nernst potential E_x of the type of ions:

$$I_x = g_x \left(\prod_i \gamma_i \right) [V_m - E_x]. \tag{3.2}$$

The gating variable γ_i, whose values range between 0 (closed) and 1 (open), is calculated using following ODE:

$$\frac{d\gamma_i}{dt} = \alpha_{\gamma_i}(1 - \gamma_i) - \beta_{\gamma_i}\gamma_i, \tag{3.3}$$

where α_{γ_i} is the forward rate constant describing the transition from closed to open state and β_{γ_i} is the backward rate constant describing the transition in the opposite direction. The rate constants depend on the transmembrane voltage V_m. Depending on the protein structure of the ion channel, one or more gating variables govern the opening, closing and inactivation of the channel. Further information on the mathematical description of myocyte membranes is given in e.g. [27, 114].

3.1.1 Models of Human Atrial Electrophysiology

One part of this thesis is the benchmark of models of human atrial electrophysiology. In the recent years, five models with diverse current formula-

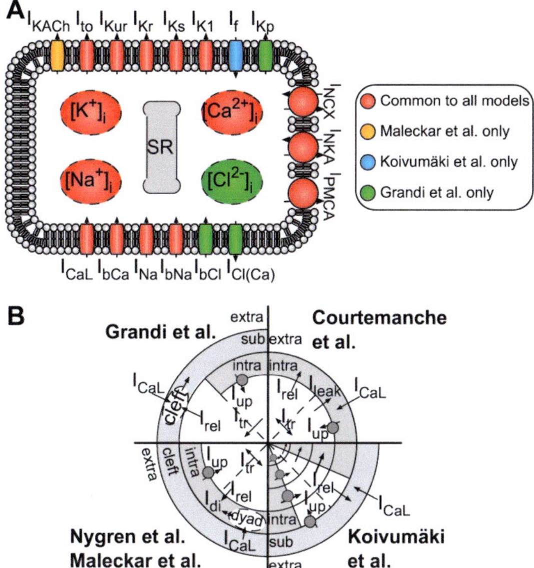

Figure 3.2. Models of human atrial myocyte according to Courtemanche et al. [115], Nygren et al. [116], Maleckar et al. [117], Koivumäki et al. [118] and Grandi et al. [119]. A: Schematic of the myocyte membrane including the different ion channels, transporters and pumps of the five models. B: Schematic description of the calcium handling in the different compartments of the five models. Modified from [13].

tions and therefore different resulting characteristics have been published. A schematic illustration of the membrane including the ion currents described by the different models is presented in Fig. 3.2A. Furthermore, the calcium handling of the models including the different compartments is shown in Fig. 3.2B. Short descriptions of the origins, similarities and differences of the models are given in the following, as well as in [13, 120].

3.1.1.1 Courtemanche et al. 1998

The model of Courtemanche et al. [115], which is termed Courtemanche model in the following, was published in 1998 as one of the first models of human atrial electrophysiology. This model was intended for the investigation of the inhibition of I_{CaL} and I_{NCX}, as well as the adaptation of the human atrial AP to changes of the heart rate. The Courtemanche model formulations of I_{Na}, I_{Kur}, I_{to}, I_{CaL}, I_{Kr}, I_{Ks} and I_{K1} are based on human experimental data. In contrast, I_{NCX}, I_{NKA}, the plasma membrane calcium ATPase current I_{PMCA}, the background calcium and sodium currents I_{bCa} and I_{bNa}, as well as the intracellular calcium handling build on the model of guinea pig ventricular myocytes of Luo and Rudy [121]. The Courtemanche model comprises intra- and extracellular compartments for the different ions. The intracellular ion concentrations are computed dynamically, whereas the extracellular concentrations are constant. The calcium uptake into the network SR is maintained by the uptake calcium current I_{up}, whereas a calcium leak current I_{leak} decreases the amount of calcium in the network SR. The junctional SR and network SR are connected by the calcium transfer current I_{tr}. The calcium release current I_{rel} supplies the intracellular space with calcium stored in the SR. Mathematical descriptions of calsequestrin, calmodulin and troponin allow for buffering of $[Ca^{2+}]_i$.

3.1.1.2 Nygren et al. 1998

The model of Nygren et al. [116] was also published in 1998. The main purpose of the Nygren model was to compare the differences between rabbit and human atrial electrophysiology, especially regarding the repolarizing currents. The Nygren model, which is mostly based on the rabbit atrial model of Lindblad et al. [122], reflects nearly the same human experimental data as in case of the Courtemanche model. In general, the Nygren and Courtemanche model present the same transmembrane ion currents; only the ultra-rapid delayed rectifier potassium current I_{Kur} is termed sustained outward potassium current I_{sus} in the Nygren model. Long term stability of the ion concentrations is achieved by an electroneutral sodium influx. Furthermore, the ions can accumulate or deplete in a cleft space between the intra- and extracellular space around the cell membrane. In case of calcium ions, an additional dyadic space between the L-type calcium channels and the junctional SR is included. At this dyadic space, the calcium concentration can be significantly increased compared to the concentration in the remaining intracellular space. A calcium diffusion current I_{di} flows from the dyadic to the intracellular space. In contrast to the Courtemanche model, which offers a phenomenological voltage-dependent calcium release, a calcium-induced calcium release is described in the Nygren model.

3.1.1.3 Maleckar et al. 2008

Ten years after the publication of the first models of human atrial electrophysiology, Maleckar et al. [117] presented a re-implementation of the Nygren model. The model was used to study the coupling between myocytes and fibroblasts and to improve the description of the rate-dependent repolarization of human atrial APs. For this purpose, more recent human atrial experimental data of the repolarizing currents I_{Kur} and I_{to} are included. The effects of cholinergic stimulation on human atrial electrophysiology are considered by adding a mathematical description of I_{KACh} according to Kneller et al. [123].

3.1.1.4 Koivumäki et al. 2011

In 2011, the model of Koivumäki et al. [118] was published. As the Maleckar model, it is based on the Nygren model. The Koivumäki model aimed at improving the description of the intracellular calcium handling and to investigate the impact of the calcium dynamics on the shape of the AP. For this purpose, three intracellular and one subspace calcium, as well as four SR compartments were defined to model spatially heterogeneous intracellular calcium dynamics. Calcium flows between the SR compartment closest to the membrane and the subspace, as well as between the three other SR compartments and the corresponding intracellular compartments. In this way, calcium diffusion due to the lack of a transverse tubular system is mathematically described. The calcium release to the intracellular compartments is modeled by a Hodgkin-Huxley type current formulation. In addition, a hyperpolarization-activated inward potassium current I_f is included according to [124].

3.1.1.5 Grandi et al. 2011

The model of human atrial electrophysiology of Grandi et al. [119] was published in the same year as the Koivumäki model. The Grandi model is based on neither the Courtemanche nor the Nygren model, but on a model of human ventricular electrophysiology published by the same group [125]. A rabbit ventricular model [126] in turn forms the basis of the human ventricular model. The focus of the human atrial model is on the investigation of

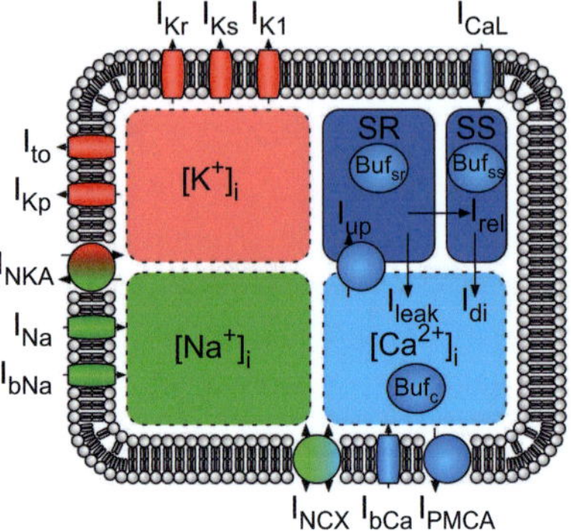

Figure 3.3. Schematic of human ventricular myocyte membrane based on the model of ten Tusscher and Panfilov [127]. Shown are the different ion channels, transporters and pumps of the membrane, as well as the calcium handling including cytosolic, SR and subspace (SS) buffers (Buf).

the differences between human atrial and ventricular electrophysiology, especially regarding the intracellular calcium handling. For this purpose, the detailed descriptions of the intracellular calcium dynamics of the rabbit ventricular model [126] were adopted. A junctional cleft between the membrane and the release unit of the SR is considered to allow for local accumulation of calcium. Furthermore, a subsarcolemmal space adjacent to the membrane is included. The concentration of chloride and therefore a calcium-activated and a background chloride current $I_{Cl(Ca)}$ and I_{bCl} are considered in addition, as well as a plateau potassium current I_{Kp}.

3.1.2 Model of Human Ventricular Electrophysiology

For the simulation of human ventricular electrophysiology, the model of ten Tusscher and Panfilov [127] published in 2006, which is called ten Tusscher model in the following, was used in this thesis. A schematic of the ventricular myocyte described by this model including the ion currents, ion compartments and the SR, is shown in Fig. 3.3. This model, which is an extension of the previous model published by the same group [128], is based on mostly human experimental data. The original purpose of the ten Tusscher model is to investigate dynamic instabilities in the APD resulting in steep restitution curves and reentrant circuits in 2D patches. The ten Tusscher model has an improved description of intracellular calcium dynamics including a Markov model of the ryanodine receptor based on [126, 129] to simulate calcium-induced calcium release. Furthermore, different buffers in the intracellular space, the SR and the subspace (SS) regulate the calcium concentrations in the different compartments of the model. Transmural heterogeneities in the human ventricles are considered by different maximal ion channel conductivities and current formulations of I_{to} and I_{Ks} for endocardial, M and epicardial myocytes.

3.1.3 Adaptation of Models of Cardiac Electrophysiology

The models of human cardiac electrophysiology introduced so far, generally describe normal physiological conditions. However, current research focuses on the investigation of other conditions, i.e. the influence of pathological changes or pharmacological treatment. In this thesis, the impact of AF, acute cardiac ischemia and pharmacological treatment on cardiac electrophysiology was investigated. Therefore, exemplary adaptations of models of cardiac physiology to these conditions are outlined in the following.

3.1.3.1 Models of Atrial Fibrillation

The Courtemanche model was adapted to cAF-induced electrical remodeling in e.g. [130, 131], where the conductivities of I_{to}, I_{Kur} and I_{CaL} were reduced to investigate the resulting effects, as e.g. AP shortening at different rates. Furthermore, the influence of pharmacological treatment on single cells, which was modeled by additional reduction of ion channel conductivities, was investigated in [130]. In [132], electrical remodeling was reproduced by reduction of the maximal channel conductivities of I_{to}, I_{Kur}, I_{CaL} and I_{K1}. In this case, the inhibition of I_{Kur} by modification of the channel kinetics was simulated to analyze the effects of this pharmacological therapy in atrial myocytes. Two descriptions of cAF-induced remodeling were investigated in [133], where reductions of maximal channel conductivities, as well as shifts of the voltage-dependent activation of I_{to}, I_{CaL}, I_{Na} and I_{K1} were considered. The impact of the different changes on human atrial APs was investigated in single myocytes. In a more recent work [134], a stretch-activated channel was included in the Courtemanche model. Under these conditions, atrial dilation facilitated the initiation and increased the stability of AF in a 3D model of the atria. The Nygren model was also adapted to reproduce cAF-induced remodeling in [131, 133], where the same changes as in the Courtemanche model were investigated. The Grandi model presents an own model of cAF-induced remodeling [119], in which the maximal channel conductivities of I_{K1}, I_{to}, I_{Kur}, I_{Ks}, I_{CaL}, I_{Na} and I_{NCX} were altered. Additionally, the intracellular calcium handling was modified by an increase of I_{leak} and an increase of the sensitivity of I_{rel} to changes of the calcium concentration in the SR. In this way, the effects of cAF on the simulated AP and the intracellular calcium dynamics were investigated. An overview of electrophysiological models of AF is given in [135].

The impact of mutations of cardiac ion channels associated with AF was investigated in e.g. [136]. In this case, the maximal channel conductivity of I_{Kur} of the Courtemanche model was reduced to describe the E375X loss-of-function mutation in KCNA5. As a consequence, early afterdepolarizations could be observed in single cells, which was assumed to be the underlying mechanism responsible for AF in case of this mutation. In [137], a voltage-dependent change of the maximal channel conductivity of I_{K1} adapted to measured I-V curves was used to simulate the effects of the V93I gain-in-function mutation in KCNJ2. The mutation caused an increased stability of reentrant circuits in 2D and 3D atrial models.

3.1.3.2 Models of Acute Cardiac Ischemia

The main effects of acute cardiac ischemia on ventricular electrophysiology were investigated in e.g. [138]. There, the impact of the transmural heterogeneity of an ATP sensitive potassium current $I_{K,ATP}$ was studied in a model of rabbit ventricular myocytes [139] after 20 min of ischemia. The focus of this study was on the metabolic effects. These effects were also investigated in detail in [140] in a model of guinea pig ventricular myocytes [121] after 15 min of ischemia. In e.g. [141–143], the focus was on phase 1b of ischemia, which was investigated in a model of guinea pig ventricular myocytes [144]. In [145], the effects of different stages of acute cardiac ischemia were analyzed in a tissue patch using a model of rabbit ventricular myocytes [146].

In [147], the effects of acute cardiac ischemia were integrated into a model of human ventricular myocytes [128]. However, only exemplary changes were included, as e.g. an additional $I_{K,ATP}$ and modifications of I_{K1}, I_{CaL}, I_{rel} and I_{up}. The effects of phase 1a of acute cardiac ischemia were integrated into the ten Tusscher model [127] of human ventricular electrophysiology in [148]. For this purpose, a current formulation of $I_{K,ATP}$ according to [149] was integrated into the model and several model parameters were adapted to dynamically model the effects at different stages of ischemia. Furthermore, the effects of ischemia, which were modeled heterogeneously, were investigated in body surface ECGs.

3.1.3.3 Models of Pharmacological Therapy

The impact of pharmacological treatment on cardiac electrophysiology was studied in e.g. [132], where the inhibition of I_{Kur} was investigated in the Courtemanche model. The effects of dofetilide, which inhibits hERG channels, on the dispersion of ventricular repolarization were analyzed in [150]. For this purpose, the inhibition was simulated using a model of guinea pig ventricular myocytes [144]. The influence of the same pharmacological agent on the ECG was observed in [151]. In this study, a model of human ventricular myocytes [125] and a 3D model of the body including the heart, which was stimulated apically, were used to simulate QT prolongation due to dofetilide. In [152], dose-dependent effects of pinacidil on $I_{K,ATP}$ and arrhythmogenesis were investigated in an ischemic 2D tissue patch using a model of guinea pig ventricular myocytes based on [121].

Transmurally varying effects of amiodarone and d-sotalol were studied in a 3D ventricular wedge in [153] using a model of canine ventricular myocytes [154]. In case of amiodarone, only the inhibition of I_{NaL}, I_{CaL} and I_{Ks} was considered by reduction of the maximal ion channel conductivities. In [155], the influence of dronedarone on human atrial electrophysiology was simulated using the Courtemanche model. For this purpose, the ion channel conductivities of I_{Na}, I_{Kr}, I_{Ks} and I_{CaL} were reduced. Due to this, bifurcation of the single-cell APD restitution curve was prevented and the APD was increased. General information on modeling the impact of pharmacological therapy on cardiac electrophysiology is given in e.g. [156].

3.2 Modeling Cardiac Excitation Propagation

In the previous section, modeling of cardiac electrophysiology was described at the single-cell level only. However, cardiac myocytes are connected by gap junctions forming a syncytium, as described in section 2.2.1. Due to this, a mathematical description of the coupling between the cells is needed. Depending on the focus of the simulation, either microscopic models describing each single myocyte in detail or macroscopic models spatially averaging small parts of the tissue, can be used. In this thesis, the tissue simulations ranged up to the organ level. Therefore, a macroscopic approach was used, since the number of cardiac myocytes of only the left ventricle is in the order of 10^9 [157].

Bidomain Model

The bidomain model is a mathematical description of electrical excitation propagation and was first applied to cardiac tissue in 1978 [158, 159]. This model spatially averages the tissue, which is represented by two continuous domains: the intra- and extracellular spaces, which are both defined at every point in space. Conductivity tensors defined at each of these points give information on the fiber orientation and coupling of the averaged tissue. Except for external stimuli, currents flowing out of one domain enter the other domain through the membrane [160]. As a consequence, the electric potential in each domain can be determined using the following Poisson equations:

$$\nabla \cdot (\sigma_i \nabla \Phi_i) = \beta I_m - I_{si} \tag{3.4}$$

$$\nabla \cdot (\sigma_e \nabla \Phi_e) = -\beta I_m - I_{se}, \tag{3.5}$$

where σ_i and σ_e are the volume-averaged conductivity tensors of the intra- and extracellular spaces, β is the membrane surface to cell volume ratio, I_m the transmembrane current density and I_{si}, as well as I_{se} are external stimulus current sources [161]. Since cardiac myocytes have approximately cylindrical shapes, the conductivity tensors can be transformed into longitudinal and transversal components, which reflect the main axes of the cells. The bidomain equations can be derived from equations (3.4) and (3.5):

$$\nabla \cdot ((\sigma_i + \sigma_e)\nabla \Phi_e) = -\nabla \cdot (\sigma_i \nabla V_m) \tag{3.6}$$

$$\nabla \cdot (\sigma_i \nabla V_m) + \nabla \cdot (\sigma_i \nabla \Phi_e) = \beta (C_m \frac{dV_m}{dt} + I_{ion}) - I_{si}. \tag{3.7}$$

To obtain equation (3.6), which is an elliptic partial differential equation, equations (3.4) and (3.5) are added and the definition of V_m is applied. Furthermore, the assumption that no external stimulus currents are applied, is necessary. The definition of the transmembrane current $I_m = C_m \frac{dV_m}{dt} + I_{ion}$ has to be used to derive the parabolic differential equation (3.7) from equation (3.4). Dirichlet and Neumann boundary conditions are applied to define the reference potential and the interface between the tissue and the surrounding medium. More detailed information on the bidomain model is given in e.g. [161, 162].

Monodomain Model

The monodomain model is described by the reaction-diffusion equation (3.8) consisting of a parabolic partial differential equation coupled to a system of ODEs. The monodomain equation can be derived from the bidomain model assuming equal anisotropy, i.e. a scalar constant κ satisfies the following equation $\sigma_i = \kappa \sigma_e$:

$$\nabla \cdot (\sigma_i \nabla V_m) = I_{inter} = (\kappa + 1)\beta \left(C_m \frac{dV_m}{dt} + I_{ion} \right). \tag{3.8}$$

Depending on the transmembrane voltage distribution, the intercellular stimulus current I_{inter} is described. If no external electrical fields are applied, the monodomain model is sufficient to investigate cardiac excitation propagation [135]. The monodomain model was used for all tissue simulations presented in this thesis, since this model is much less computationally expensive than the bidomain model.

3.3 Anatomical Models and the Forward Problem of Electrocardiography

An anatomical voxel model (cubic voxel size 0.4 mm) of the ventricles, which was derived from magnetic resonance images of a healthy volunteer

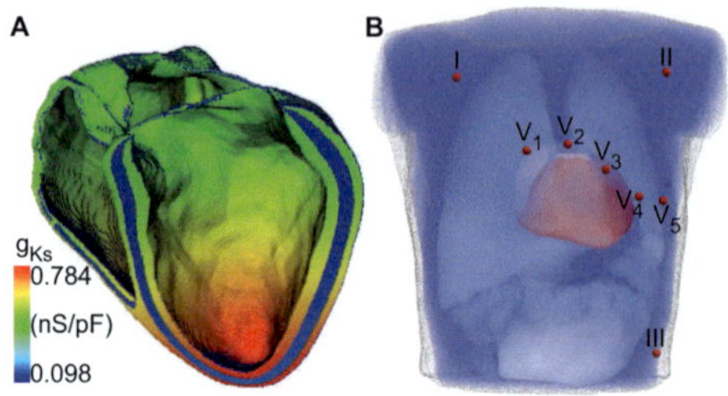

Figure 3.4. Anatomical models for whole organ and body modeling. Heterogeneous ventricular (A) and torso model (B) of healthy volunteer used for forward calculation of body surface potential maps (BSPMs). A: transmural and apico-basal heterogeneity of g_{Ks}. B: torso model with virtual ECG electrodes (I–III: Einthoven leads, V_1–V_5: Wilson chest leads, V_6 on the back) and the internal organs.

in a previous work [163], was used in this thesis for whole organ modeling. As described in [163], the transmurally varying tissue conductivity was scaled to fit canine ventricular wedge experiments [164] resulting in a longitudinal CV of approximately 650 mm/s. The same endocardial stimulation profile mimicking the His-Purkinje network and the fiber orientation described in [163] were used. For the simulation of ventricular excitation propagation using the ten Tusscher model described in section 3.1.2, different heterogeneities were considered. For this purpose, the transmural tissue layers were distributed as follows: 40% endocardial, 40% midmyocardial and 20% epicardial tissue. Consequently, the transmural heterogeneities of I_{Ks} and I_{to} suggested in [127] were applied. Furthermore, the values of the maximal channel conductivity g_{Ks} were linearly increased from the base to the apex to the twofold value according to [165]. The resulting ventricular model including the different heterogeneities of g_{Ks} is shown in Fig. 3.4A.

For the forward calculation of body surface potentials, a high resolution tetrahedral model of the torso consisting of approximately 0.75 million nodes, which was also derived from magnetic resonance images of the same healthy volunteer [163], was used. As outlined in [163], the transmembrane voltage distributions resulting from simulations of ventricular excitation propagation were interpolated on the tetrahedral mesh first. Then, the body surface potentials were computed at each simulated millisecond using equation (3.6). This equation was transformed into a linear system of equations using the finite element method and afterwards solved with the help of the Cholesky decomposition and a conjugate gradient method. The torso including the internal organs, as well as the virtual electrodes of the 12-lead ECG are presented in Fig. 3.4B. More information on the forward calculation of body surface potentials is given in e.g. [166].

Part II

Methods

4

Parameter Adaptation of Electrophysiological Models

This chapter describes the methods necessary for the adaptation of models of cardiac electrophysiology to voltage or patch clamp measurement data, which were introduced in section 2.2.2. In this study, different optimization algorithms for the adjustment of a model of I_{Kr} to voltage clamp measurement data of the hERG channel, which is the α-subunit of I_{Kr}, were investigated. At first, the experimentally measured and synthetic data used for the test of the optimization algorithms are described. Then, the necessary preprocessing of the noisy measurement data of single myocytes is presented. Furthermore, two methods to solve ODEs modeling the gating processes of cardiac ion channels are compared. Three algorithms, which pursue different strategies to minimize the difference between measured and simulated current traces, were analyzed in this work. Finally, a local sensitivity analysis of the influence of the adjustable parameters on the function to be minimized was carried out. The corresponding results can be found in chapter 8.

4.1 Measured hERG and Synthetic Voltage Clamp Data

In this study, exemplary hERG double microelectrode voltage clamp data were used for the test of optimization algorithms. However, patch clamp measurement data or data of other currents can be used instead, as shown e.g. in section 5.3. For the test of the optimization algorithms, currents of wild type (WT) hERG and two mutations of hERG, namely K897T and N588K, were provided by the University Hospital Heidelberg. The *QuickChange Site-Directed Mutagenesis Kit (Stratagene, La Jolla, CA, USA)* was used to introduce both mutations into the hERG cDNA. 46 nl cRNA solution were injected in each Xenopus oocyte to heterologously express hERG channels there. The oocytes were incubated at 16°C for three to four days. Then, the voltage clamp recordings were performed using a *Warner OC-725A* amplifier (*Warner Instruments, Hamden, CT, USA*) at 23 to 25°C. The microelectrodes had tip resistances in the range of 1 to 5 MΩ. The currents were low-pass filtered at 1 to 2 kHz and digitized at 5 to 10 kHz (*Digidata 1322A, Axon Instruments, Union City, CA, USA*). The bathing solution contained 5 mM KCl, 100 mM NaCl, 1.5 mM CaCl$_2$, 2 mM MgCl$_2$, and 10 mM of the buffering agent HEPES. The pH was adjusted to 7.4 with NaOH. The pipette solution contained 3 M KCl. The data of the WT channels were used to adapt and compare the optimization algorithms. The algorithms were furthermore tested with the voltage clamp data of the mutated channels.

In addition to the experimentally measured hERG currents, synthetic current traces were used to assess the ability of the optimization algorithms to reconstruct known parameter values. For this purpose, current traces were obtained from simulation of the same voltage clamp protocol as used for the measured data. The I_{Kr} current formulation of the Courtemanche model [115] with the 12 adjustable parameters (*grey*) is:

$$I_{Kr} = \frac{g_{Kr} x_r (V_m - E_K)}{1 + exp\left(\frac{V_m + g_{Kr1}}{g_{Kr2}}\right)} \tag{4.1}$$

$$\alpha_{x(r)} = Xr_{a1}\frac{V_m + Xr_{a2}}{1 - exp\left(\frac{V_m + Xr_{a2}}{Xr_{a3}}\right)} \tag{4.2}$$

$$\beta_{x(r)} = 7.3898e - 5\frac{V_m + Xr_{b1}}{exp\left(\frac{V_m + Xr_{b1}}{Xr_{b2}}\right) - 1} \tag{4.3}$$

$$\tau_{x(r)} = [Xr_{KQ10}(\alpha_{x(r)} + \beta_x)]^{-1} \tag{4.4}$$

$$x_{r(\infty)} = \left[1 + exp\left(-\frac{V_m + Xr_{m1}}{Xr_{m2}}\right)\right]^{-1} \tag{4.5}$$

$$E_K = \frac{RT}{F}log\frac{[K^+]_o}{[K^+]_i}, \tag{4.6}$$

where g_{Kr} was the maximum conductivity and Xr_{KQ10} the temperature correction factor. The Nernst voltage E_K based on the gas constant R (8.3143 J/(K · mol)), the temperature T (297.15 K) and Faraday's constant F (96.4867 C/mmol) was computed with constant intracellular potassium concentration $[K^+]_i$.

The gating variable x_r was determined using the ODE described in equation (4.7). The original values of the adjustable parameters (compare Table 8.1) of the model were used for the generation of synthetic clamp data. Then, the optimization algorithms, which received random initial parameter values as an input, were supposed to reconstruct the original parameters again.

All current traces, i.e. the measured and synthetic data, and the corresponding voltage clamp protocol are depicted in Fig. 8.7.

4.2 Preprocessing of Measured Ion Current Recordings

As the currents recorded during voltage or patch clamp protocols are noisy and relatively small (e.g. in the range of μA) and vary from cell to cell, the measurements of several cells were averaged first. Furthermore, the aim of the adaptation of ion current models to measurement data was not to model the properties of a specific single cell, but rather to reproduce the characteristics of a cohort of certain cells. For this purpose, the current traces of all cells were normalized to their maximal peaks (positive or negative) first, since the amplitudes varied from cell to cell depending on the size and expression of the ion channels. Then, the normalized current traces of different cells corresponding to the same voltage step were averaged. Afterwards, each of these averaged current traces was multiplied by the average of the maximal peaks of the different cells. In this way, information on the amplitude of the current traces corresponding to different voltage steps was included again. In addition, the averaging reduced the remaining high-frequency noise.

The cells, which were used for voltage clamp or patch clamp measurements in this work, were injected with the cRNA encoding the ion channel that should be investigated. In this way, the cells mainly expressed the channel of interest. Nevertheless, background and leak currents of the cells also contributed to the measured currents. As a consequence, the measured current traces were slightly shifted, which could be identified in parts of the voltage clamp protocol, during which the current of interest was supposed to be zero. If available, the current recordings were corrected with measurements of the background currents using cells without injected cRNA. These measurements, which were also averaged as described above, were scaled, so that the difference between the current of interest and the background current was zero at clamp voltages, at which the channels of interest were supposed to be closed, e.g. at $-80\,\mathrm{mV}$. If these additional recordings were not available, the shifts of the current traces were corrected manually, so that the current traces were zero on average, if the channels were supposed to be closed.

Finally, the number of measurement points was decreased to reduce computing time of the optimization process, which depended on the number of function evaluations calculating the difference between simulated and measured data. Additionally, points immediately after instant changes of the clamp voltage were discarded, as the measured current could be influenced by the capacitive properties of the cell membrane.

4.3 Methods for the Solution of ODEs

ODE (3.3) describing the gating process of cardiac ion channels can be rewritten as follows:

$$\frac{d\gamma}{dt} = \frac{\gamma_\infty - \gamma}{\tau_\gamma}, \tag{4.7}$$

where γ_∞ is the steady state value and τ_γ the time constant of the gating process. Both, γ_∞ and τ_γ depend on the transmembrane voltage V_m. In general, two possible methods to solve these ODEs in *Matlab (R2012b, The MathWorks, Natick, MA, USA)* were compared in this work.

Analytical Solution

As the transmembrane voltage was a piecewise constant function during the voltage clamp protocols used in this work, the values of γ_∞ and τ_γ were also constant during defined intervals, except for the instant changes in V_m. Consequently, an analytical solution of equation (4.7) existed, as shown in e.g. [167]:

$$\gamma(t - t_0) = \gamma_\infty + (\gamma_0 - \gamma_\infty)\exp\left(-\frac{(t - t_0)}{\tau_\gamma}\right), \tag{4.8}$$

with t_0 being the time of a voltage step and γ_0 the corresponding initial value at t_0.

Numerical Solution

In computational cardiology, ODE (4.7) is not solved directly, but approximated numerically, since the transmembrane voltage and therefore γ_∞ and τ_γ generally change over time during the cardiac cycle. The predefined solver *ode15s*, which numerically approximates stiff ODEs with a variable time step, was used for the optimizations. In order to obtain an approximation with sufficient accuracy, the error tolerance was set to $1e-9$.

4.4 Optimization Algorithms

In this work, the parameters of the models of cardiac ion currents were adapted to reproduce the properties of measured ion currents. For this purpose, the difference between simulated and measured ion currents was minimized. The quality of the adaptation was quantified using the least squares:

$$\min_{\mathbf{p}} \left(\sum_j \sum_i \left(I_{Kr,meas}(t_i, V_j) - I_{Kr,sim}(t_i, V_j, \mathbf{p}) \right)^2 \right), \tag{4.9}$$

with $\mathbf{p}$ being the set of adjustable parameters, t_i discrete evaluation times and V_j the clamp voltages. V_j was described by a piecewise constant function reproducing the voltage clamp protocol of the measurements. The resulting optimization problem was nonlinear, multi-dimensional and non-convex. Therefore, different optimization algorithms were investigated regarding their suitability for this task.

4.4.1 Trust-Region-Reflective Algorithm

The trust-region-reflective (TRR) algorithm used in this work is implemented in the *Matlab* function *lsqnonlin*. Instead of minimizing the nonlinear function directly, a quadratic approximation $q_k(x)$ of the function $f(x_k)$, which is defined in a trust region with radius r_Δ around the parameter vector x_k, is minimized [168]:

$$q_k(x) = f(x_k) + \nabla f(x_k)^T (x - x_k) + \tfrac{1}{2}(x - x_k)^T \nabla^2 f(x_k)(x - x_k) \tag{4.10}$$

$$\min_{\|s\| \leq r_\Delta} q_k(x_k + s), \tag{4.11}$$

with s being the difference between x_k and x. In the course of the optimization, the trust region radius Δ and the parameter vector x_k are adapted. In this work, the maximum number of iterations was set to 1e5 and the number of maximal function evaluations to 2.5e6 to restrict the maximum computing time of this algorithm. As a further stopping criterion of the algorithm, the minimal changes of the parameter and function values were set to $1e-11$. However, the results of the TRR algorithm can depend on the initial parameter vector, if it is applied to problems with several local minima [168].

4.4.2 Derivative-Free Optimization Algorithms

In addition to the TRR algorithm, which was based on information about the derivatives, a different class of heuristics exists, which does not consider this information. Instead, stochastic optimization algorithms randomize the search process to avoid running into local minima, trying to find a global minimum. Both derivative-free algorithms implemented in *Matlab* in this study use a population of parameter vectors instead of only one. The computation of the iterations of individuals of these populations was parallelized.

4.4.2.1 Particle Swarm Optimization

The particle swarm optimization (PSO) algorithm is based on swarming theory of animals, such as flocking birds [169]. The algorithm describes the behavior of a population of particles starting from random initial positions, which correspond to the parameter vectors, within certain boundaries. The entire population tries to move towards the global best position. For each particle, the velocity $\mathbf{v}_i$ depending on the current global best postion $\mathbf{p}_g$ and the own best position $\mathbf{p}_i$ is computed in every iteration as described e.g. in [170]:

$$\mathbf{v}_i \leftarrow \chi(\mathbf{v}_i + \mathbf{U}(0,\phi_1) \otimes (\mathbf{p}_i - \mathbf{x}_i) + \mathbf{U}(0,\phi_2) \otimes (\mathbf{p}_g - \mathbf{x}_i)), \qquad (4.12)$$

$$\mathbf{x}_i \leftarrow \mathbf{x}_i + \mathbf{v}_i, \qquad (4.13)$$

with $\mathbf{U}(0,\phi_1)$ and $\mathbf{U}(0,\phi_2)$ being vectors of uniformly distributed random numbers between 0 and ϕ_1 or ϕ_2. $\otimes$ is a component-wise multiplication. The values of $\mathbf{p}_g$ and $\mathbf{p}_i$ are updated, as soon as better positions are found in an iteration. The constriction coefficient χ was defined by Clerc and Kennedy [171] as:

$$\chi = \frac{2}{\phi - 2 + \sqrt{\phi^2 - 4\phi}}, \text{ with } \phi = \phi_1 + \phi_2 = 4.1 \Rightarrow \chi \approx 0.7298. \quad (4.14)$$

The weighting factors ϕ_1 and ϕ_2 were equal in this case. Furthermore, the resulting parameter values were set to values within the boundaries, if they were determined to be out of the defined range. For this purpose, the corrected parameter value was placed within 25% of the defined range starting from the boundary. If the number of particles N was sufficiently large, the swarm was supposed to move to the global best position avoiding local minima. In contrast to the previous algorithms, a fixed number of iterations was used.

4.4.2.2 Genetic Algorithm

Genetic algorithms (GA) are a further class of derivative-free optimization heuristics using a population of N parameter sets. As in case of the PSO algorithm, the GA implemented in this work had a fixed number of iterations. The heuristic was based on three steps inspired by biological evolution:

1. *Selection:* The function to be minimized was evaluated for all parameter sets in parallel. The n_{alpha} best parameter sets of a generation, which were called alphas as in [172], were selected first. Then, the $N - n_{alpha}$ other parameter sets were divided into groups of $n_{alpha}/N - 1$ remaining parameter sets, which were assigned to one alpha each.
2. *Crossover:* In this step, the parameter values of the remaining parameter sets were crossed over with the values of their corresponding alpha. The probability $p_{crossover}$ of each parameter value of the remaining parameter sets to inherit the value of the alpha was independent of those of the other parameters.
3. *Mutation:* Finally, all parameter values of the alphas and remaining parameter sets were mutated with a probability $p_{mutation}$. The range of the perturbation of the parameter value was decreased with increasing number of iterations as described in [173]. In the first iteration, the perturbations ranged from $\pm 50\%$ of the maximum difference between the upper and lower boundaries and the current value. In the last iterations, the range centered around the current value was decreased to $\pm 2.5\%$, so that

the minimum search was focused on more local parameter ranges towards late iterations.

4.5 Local Sensitivity Analysis

The aim of a sensitivity analysis is to quantify, how sensitive an obtained solution is to variations of one or more input parameters. In this work, the local sensitivity S based on the partial derivatives, i.e. the influence of changes of one parameter at a time on the solution, was analyzed:

$$S_i = \frac{df(x_i)}{dx_i}.$$
(4.15)

As shown in equation (4.8), an analytical solution of the ODEs of the models of cardiac ion currents existed in case of piecewise constant transmembrane voltages during voltage clamp experiments. Due to this, the partial derivatives of the function to be minimized in equation (4.9) with respect to the adjustable parameters could be computed directly using *Maple (v. 15, 2011, Maplesoft, Waterloo, ON, Canada)*. Therefore, the sensitivity of the minimum, which was found using the optimization algorithms, to changes of the adjustable parameters was investigated.

In contrast to local sensitivity, global sensitivity, which is based on e.g. the variance, analyzes how changes of combinations of parameters affect the solution. An overview of the topic sensitivity analysis is given e.g. in [174].

5

Modeling Atrial Fibrillation

The methods used for multiscale simulations of AF effects, which are summarized in section 2.3, are described in this chapter. Five models of human atrial electrophysiology were benchmarked regarding general model properties and their ability to reproduce effects of cAF-induced remodeling. Furthermore, the impact of mutations of the ultra-rapid delayed rectifier potassium current integrated into a model of human atrial electrophysiology with methods described in chapter 4 were investigated. The scale of the simulations ranged from single myocytes up to 2D tissue patches. The initiation of reentrant circuits was analyzed in these patches. For this purpose, the rotor centers were detected automatically and pseudo-ECGs were computed in order to determine the dominant frequency. The corresponding results are presented in chapter 9.

5.1 General Properties of Models of Atrial Electrophysiology

The principal properties of the five models of human atrial electrophysiology presented in section 3.1.1 (Courtemanche, Nygren, Maleckar, Koivumäki and Grandi) were investigated first in order to determine, which of them was most suitable for the simulation of effects of AF. For this purpose, some aspects that are important for the simulation of human atrial electrophysiology were benchmarked using the models as they were published:

Long Term Stability

In order to assess the numerical long term stability of the models, some AP parameters of interest were observed over long simulation times. For this purpose, the resting membrane voltage $V_{m,rest}$ was recorded during a period of 20 min without external stimulation, so that $V_{m,rest}$ did not change significantly. After this quiescent phase, the myocytes were paced with a BCL of 1 s and $V_{m,rest}$ was measured for each cycle. In this way, the reaction of the models to onset of external stimulation was examined. Furthermore, the AP amplitude, APD_{50} and APD_{90} (AP durations at 50% and 90% repolarization) of each cycle were analyzed during 20 min pacing with a BCL of 1 s.

Alternans

Alternans is a physiological phenomenon, which occurs if the BCL is only slightly higher than the APD, so that the myocytes are not able to recover completely from the previous beat. Consequently, a smaller AP with shorter duration is initiated in the following causing a longer DI than before. Due to this, a longer AP is initiated again, which results in an alternating pattern of short and long APs. This behavior was investigated in the five models of atrial electrophysiology by pacing the myocytes for 30 s with BCLs ranging from 0.2 to 1 s. The APD_{50} of different numbers of stimuli was measured to investigate alternating behavior of the APD. Furthermore, the resulting APs and the calcium transient were observed during 30 s pacing with a BCL of 0.25 s.

Computing Time

Another aspect important for the simulation of human atrial electrophysiology, e.g. fibrillation in a complex geometry, is the computing time. The more complex the models are, the more time is needed for the solution of the underlying mathematical equations. The single-cell computing times of the different models were determined by simulation of 1000 s with a BCL of 1 s on Mac OSX 10.7 (*Apple Inc., Cupertino, CA*) with a 2.7 GHz Intel Core i7 and 8 GB RAM. In this case, no results were written to hard disk to avoid unwanted delays.

5.2 Model of Electrical Remodeling due to Chronic AF

As described in section 2.3.1, processes changing the expression and functionality of ion channels and gap junctions occur during cAF. This is called electrical remodeling. Table 5.1 summarizes the changes of ion channel conductivities, which were considered in this work. These changes were inte-

Table 5.1. Change of ion channel conductivities (% of original value) describing electrical remodeling due to cAF.

conductivity	g_{to}	g_{Kur}	g_{CaL}	g_{K1}
change	-65%	-49%	-65%	$+110\%$
source	[46]	[46]	[45]	[47]

grated into the five models of human atrial electrophysiology in order to assess, which of them is most suitable for the simulation of cAF. Furthermore, gap junctional remodeling was considered by reduction of the intracellular tissue conductivity by 30% in tissue simulations of cAF as presented in [175].

In order to investigate the effects of cAF, various electrophysiological properties were analyzed. After 50 s paced with a BCL of 1 s, the AP amplitude and durations (APD_{50} and APD_{90}), maximum upstroke velocity dV/dt_{max}, as well as the resting membrane voltage $V_{m,rest}$ of the control and cAF models were compared. The stimulus amplitudes were adapted model specific, so that they were twice the threshold amplitude, which was needed to initiate APs. Additionally, the restitution curves of the APD_{90}, ERP, CV and wave length (WL, see also section 7.4.2 for further explanation) of the control and cAF models were determined in a 1D tissue strand of $200 \times 1 \times 1$ cubic voxels (size 0.1 mm). The models were paced for 50 beats in a single-cell environment with varying BCL between 0.2 and 1 s. After this phase of initialization, all single-cell variables were transferred to a tissue simulation environment. Then, five beats were computed in tissue with stimulus amplitudes 20% above threshold amplitude. Furthermore, model-specific values for the isotropic tissue conductivity were defined, so that a CV of approximately 750 mm/s was obtained in the control case at a BCL of 1 s. The restitution properties were evaluated at the last beat in tissue and only, if all previous beats initiated APs [13].

5.3 Modeling Genetic Defects Favoring AF

Section 2.3.2 described that genetic defects of cardiac ion channels are known to favor AF. In this work, the impact of mutations of the KCNA5 gene, which encodes I_{Kur}, was investigated. The whole-cell patch clamp recordings of WT and three mutations of this channel (A576V, E610K and T527M) known to be associated with AF [176] were provided by the University Hospital Heidelberg. The chinese hamster ovary (CHO) cells used for this purpose were cultured in minimum essential medium α in an atmosphere of 95% humidified air and 5% CO_2 at a temperature of 37°C. This medium was supplemented with $100\,\mu$g/ml streptomycin sulphate, 10% fetal bovine serum and $100\,$U/ml penicillin G sodium. Prior to treatment, the CHO cells were passaged regularly and subcultured. *Fugene (PROMEGA, Madison, USA)* transfection reagent was used for transient transfections of the CHO cells according to the manufacturer's instructions. The same amount of DNA ($3\,\mu$g per bowl) was used for this purpose. The patch clamp recordings were performed at a temperature of 37°C. The pipette solution contained $100\,$mM K aspartate, $20\,$mM KCl, $2\,$mM $MgCl_2{}^*6H_2O$, $1\,$mM $CaCl_2{}^*2H_2O$, $10\,$mM HEPES, $10\,$mM EGTA and $2\,$mM Na_2ATP. The pH value of the pipette solution was adjusted to 7.2 using KOH. The bathing solution consisted of $140\,$mM NaCl, $5\,$mM KCl, $1\,$mM $MgCl_2{}^*6H_2O$, $1.8\,$mM $CaCl_2{}^*2H_2O$, $10\,$mM HEPES and $10\,$mM glucose monohydrate. NaOH was used to adjust the pH of the bathing solution to 7.4.

The current traces of the WT and mutated channels and the corresponding clamp protocol are shown in Fig. 9.6. The adaptation of the model of I_{Kur} to the patch clamp data was carried out using the methods introduced in chapter 4.

The current formulation of the Courtemanche model with the 24 adjustable parameters (*blue, red*) was:

$$I_{Kur} = g_{Kur} u_a^3 u_i (V - E_K) \tag{5.1}$$

$$g_{Kur} = g_{Kur1} + \frac{g_{Kur2}}{1 + \exp\left(\frac{V + g_{Kur3}}{g_{Kur4}}\right)} \tag{5.2}$$

$$E_K = \frac{RT}{F} \log \frac{[K^+]_o}{[K^+]_i} \tag{5.3}$$

$$\alpha_{ua} = ua_{a1} \left[\exp\left(\frac{V + ua_{a2}}{ua_{a3}}\right) + \exp\left(\frac{V + ua_{a4}}{ua_{a5}}\right) \right]^{-1} \tag{5.4}$$

$$\beta_{ua} = 0.65 \left[ua_{b1} + \exp\left(\frac{V + ua_{b2}}{ua_{b3}}\right) \right]^{-1} \tag{5.5}$$

$$\tau_{ua} = [ua_{KQ10}(\alpha_{ua} + \beta_{ua})]^{-1} \tag{5.6}$$

$$u_{a\infty} = \left[1 + \exp\left(\frac{V + ua_{m1}}{ua_{m2}}\right)\right]^{-1} \tag{5.7}$$

$$\alpha_{ui} = \frac{ui_{a1}}{ui_{a2} + \exp\left(\frac{V + ui_{a3}}{ui_{a4}}\right)} \tag{5.8}$$

$$\beta_{ui} = \left[\exp\left(\frac{V + ui_{b1}}{ui_{b2}}\right)\right]^{-1} \tag{5.9}$$

$$\tau_{ui} = \left[ui_{KQ10}(\alpha_{ui} + \beta_{ui})\right]^{-1} \tag{5.10}$$

$$u_{i\infty} = \left[1 + \exp\left(\frac{V + ui_{m1}}{ui_{m2}}\right)\right]^{-1} \tag{5.11}$$

ua_{KQ10} and ui_{KQ10} were the temperature correction factors, the temperature T was 310 K and the intracellular potassium concentration $[K^+]_i$ was constant. The gating variables u_a and u_i were determined using the analytical solution of the ODE described in equation (4.8). After the adjustment of the parameters within defined ranges as specified in Table 5.2, the resulting parameter sets were used to calculate relative changes between the data of mutated and WT channels. For this purpose, the relative changes of the parameters p_{mod}, which appeared in a sum (*blue*) or which were multiplied/divided (*red*), were determined as follows:

$$p_{mod,sum} = (p_{mutation} - p_{WT}) + p_{control} \tag{5.12}$$

$$p_{mod,mult} = \frac{p_{mutation}}{p_{WT}} \cdot p_{control} \tag{5.13}$$

In this way, parameter changes due to the experimental settings during the patch clamp measurements were canceled out and only the changes due to the mutations remained. Consequently, three parameter sets reflecting the relative changes due to the mutations A576V, E610K and T527M were compared to the control model in multiscale simulations. As described in section 5.2, the effects of the mutations on the APs were investigated in the single cell and on the restitution properties in the 1D tissue patch.

Table 5.2. Ranges of the adjustable parameters of the ultra-rapid delayed rectifier current for the adaptation to patch clamp recordings of WT and mutated channels. For the adaptation to data of the mutated channels, the boundaries of the temperature correction factors ua_{KQ10} and ui_{KQ10} were fixed to the values resulting from the adaptation to the data of WT channels, as the same measurement conditions were assumed. The lower boundaries of g_{Kur1} and g_{Kur2} were set to the values of the WT adaptation to prevent negative conductivity values after the calculation of relative changes.

parameter	original	WT		mutations	
		lower	*upper*	*lower*	*upper*
ua_{a1}	0.65	6.5E-3	65	6.5E-3	65
ua_{a2}	10	−100	80	−100	80
ua_{a3}	−8.5	−100	−1	−100	−1
ua_{a4}	−30	−100	80	−100	80
ua_{a5}	−59	−100	−1	−100	−1
ua_{b1}	2.5	0.25	25	0.25	25
ua_{b2}	82	−60	140	−60	140
ua_{b3}	17	1	100	1	100
ua_{m1}	30.3	−100	80	−100	80
ua_{m2}	−9.6	−100	−1	−100	−1
ua_{KQ10}	3	0.01	100	29.65	29.66
ui_{a1}	1	0.01	100	0.01	100
ui_{a2}	21	0.21	2100	0.21	2100
ui_{a3}	−185	−230	−50	−230	−50
ui_{a4}	−28	−100	−1	−100	−1
ui_{b1}	−158	−200	−20	−200	−20
ui_{b2}	−16	−100	−1	−100	−1
ui_{m1}	−99.45	−150	30	−150	30
ui_{m2}	27.48	1	100	1	100
ui_{KQ10}	3	0.01	100	0.0111	0.0112
g_{Kur1}	0.005	5e-5	5E2	68.9702	5E2
g_{Kur2}	0.05	5e-4	5E3	119.757	5E3
g_{Kur3}	−15	−100	80	−100	80
g_{Kur4}	−13	−100	−1	−100	−1

5.4 Initiation of Functional Reentries in 2D Atrial Tissue Patch

For the investigation of AF in atrial tissue simulations, reentrant circuits were initiated in 2D tissue patches of $1000 \times 1000 \times 1$ cubic voxels (size 0.1 mm). For this purpose, the models were initialized for 50 beats in a single-cell environment paced at a short BCL (0.4 s). In this way, the APs and underlying processes of the models could adapt to the high pacing frequency. Then, one side of the patch was stimulated five times at the same BCL of 0.4 s. After

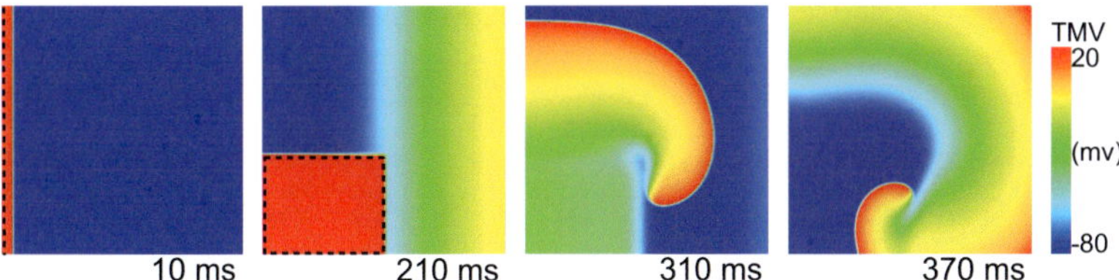

Figure 5.1. Exemplary initiation of rotor in a 2D atrial tissue patch using the electrically re-modeled Courtemanche model. At first, regular stimulus lines were placed at one side of the patch with a high BCL (e.g. 0.4 s) to create planar wave fronts. A following cross-field stimulus perpendicular to the original excitation wave was placed at the wave back ($t = 210$ ms). After 310 ms, the excitation wave followed the re-excitable tissue around the stimulated area, which was excited again in the following ($t = 370$ ms). Stimulus sites are highlighted by dashed lines. TMV: transmembrane voltage.

the initiation of the last regular excitation wave, a cross-field stimulus was placed (compare Fig. 5.1), when the wave back reached the center of the patch. In this case, the stimulated area was a rectangle reaching from the previous stimulation site to the transition of still refractory and re-excitable tissue in direction of the regular excitation wave. Perpendicular to the direction of the regular excitation wave, the stimulus was smaller than half of the width of the tissue patch, so that the new excitation wave could move in direction of the re-excitable tissue. In the exemplary cross-field (S1–S2) protocol shown in Fig. 5.1, the new excitation wave had entered the area of the cross-field stimulus after 370 ms creating a rotor.

5.5 Detection of Rotor Centers

For the analysis of reentrant circuits, an algorithm for the automatic detection of rotor centers in a 2D tissue patch was implemented. It was based on the algorithm of Bray et al. [177] from 2001. This algorithm, which can also be extended to determine filaments in 3D tissue [178], is based on the calculation of the topological charge n_t:

$$n_t = \oint_c \nabla \Theta \cdot d\mathbf{c}. \tag{5.14}$$

This is the line integral of the gradient of the phase Θ along a closed path c, as comprehensively described in e.g. [179]. If the transmembrane voltage

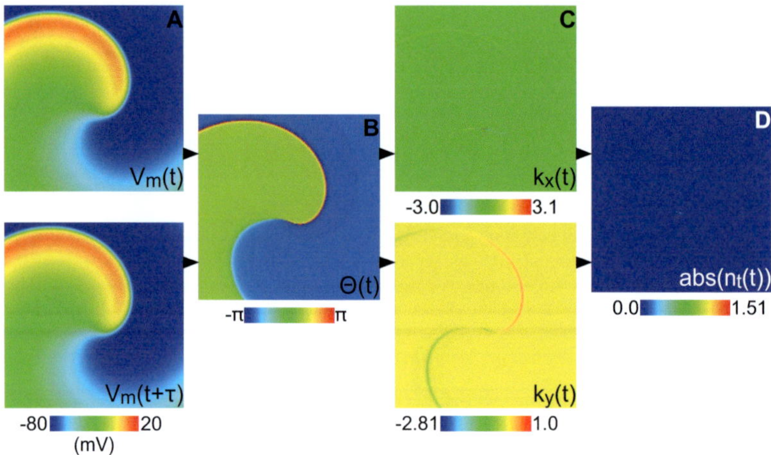

Figure 5.2. Detection of rotor centers in a 2D tissue patch. The phase Θ (B) was calculated using two consecutive transmembrane voltage V_m distributions (A). The topological charge n_t (D), which was nearly zero except for the phase singularity, was then computed as the curl of the gradient of Θ (C).

distributions are transformed into the phase space, a rotor center can be interpreted as a point, in which all phases meet, a so-called phase singularity. In this case, the phase was computed using two consecutive TMV distributions, as also exemplarily shown in Fig. 5.2:

$$\Theta(x,y,t) = \text{atan2}[V_m(x,y,t+\tau) - V_m^*, \ V_m(x,y,t) - V_m^*], \tag{5.15}$$

with τ being the time increment between two consecutive TMV distributions (typically $\tau = 1\,\text{ms}$) and $V_m^* = -40\,\text{mV}$ the origin of the phase. Since the algorithm was based on the phase, it was independent of e.g. the amplitude of the APs, which can vary depending on the model used. Then, the gradient of the discretized phase $\Theta[m,n,t]$ was approximated using forward differences $k_x[m,n,t]$ and $k_y[m,n,t]$:

$$k_x[m,n,t] = \Theta[m+1,n,t] - \Theta[m,n,t] \tag{5.16}$$
$$k_y[m,n,t] = \Theta[m,n+1,t] - \Theta[m,n,t]. \tag{5.17}$$

However, since the values of the phase gradually increased from $-\pi$ to π, phase jumps at the transition to $-\pi$ occurred, which would have been interpreted as a discontinuity in phase. Therefore, phase jumps greater than π were compensated by conversion to the 2π complement as described in [178]. Afterwards, the line integral was determined by calculation of the curl of the gradient, which was nearly zero except for the phase singularity, i.e. the rotor center. However, post-processing as mentioned in [180] was necessary, e.g. due to the discretization of the TMV distributions. This discretization was responsible for blurring of the phase singularity, i.e. the value of n_t defined by equation (5.14) gradually decreased around the phase singularity to nearly zero. Consequently, a threshold of the absolute vale of n_t was defined. The rotor center was then determined as the local maximum of the values above the threshold.

5.6 Pseudo-ECG and Dominant Frequency

As the dominant frequency (DF) is correlated with the persistency of AF, i.e. high DFs indicate more persistent forms of AF [181], this measure was determined to quantify the arrhythmic potency of the models of AF. For this purpose, pseudo-ECGs of the 2D rotor simulations were computed first. The electric potential Φ within an infinite medium with homogeneous conductivity was used as described in e.g. [182]:

$$\Phi(x',y',z',t) = \frac{1}{4\pi\sigma} \int \frac{I_v(x,y,z,t)}{r(x,y,z,x',y',z')} dxdydz, \qquad (5.18)$$

where σ is the electrical conductivity, I_v is the intercellular current density and r is the distance between the current source and a virtual measurement electrode. To obtain a pseudo-ECG, the difference between the potentials at two virtual electrodes was determined for a series of time steps t:

$$pECG = \Phi(x_1,y_1,z_1,t) - \Phi(x_2,y_2,z_2,t). \qquad (5.19)$$

The electrodes were placed 5 mm above the center of the 2D patch with a distance of 10 mm between each other. Afterwards, the power spectral density of the derived signal was computed using the fast Fourier transform for spectral analysis. The frequency with the highest peak is called DF [181].

6

Modeling Effects of Acute Cardiac Ischemia

This chapter describes the methods necessary for multiscale simulations of the effects of acute cardiac ischemia introduced in section 2.4 ranging from single cells up to the body surface ECG. The main effects of different stages of acute cardiac ischemia were integrated into an electrophysiological model of a healthy ventricular myocyte. Furthermore, regional heterogeneities depending on the occlusion site and transmurally varying ischemia effects were considered during tissue simulations. Cellular uncoupling during phase 1b was considered in the forward calculation of ECGs. The corresponding results are presented in chapter 10.

6.1 Model of a Cardiac Myocyte at Different Ischemia Stages

The effects of the first thirty minutes of acute ischemia were integrated into the electrophysiological model of ventricular myocytes of ten Tusscher and

Table 6.1. Parameter values of the model of ventricular myocytes at different stages of acute cardiac ischemia as shown in [16, 17].

model parameter		control	stage 1	stage 2	phase 1b
		0 min	*5 min*	*10 min*	*20–30 min*
$[K^+]_o$	(mmol/l)	5.4	8.7	12.5	15.0
g_{Na}, g_{CaL}	(%)	100	87.5	75	50
$dV_{m,Na}$	(mV)	0	1.7	3.4	3.4
$[ATP]_i$	(mmol/l)	6.8	5.7	4.6	3.8
$[ADP]_i$	(μmol/l)	15	87.5	99	101.5
k_{NaCa}	(%)	100	100	100	20
P_{NaK}	(%)	100	100	100	30
V_{rel}	(%)	100	100	100	5
V_{maxup}	(%)	100	100	100	90

Panfilov [127] with modifications of Weiss et al. [148]. This altered model had e.g. an additional ATP sensitive potassium channel $I_{K,ATP}$ and reproduced the main effects of phase 1a ischemia – hypoxia, hyperkalemia and acidosis – dynamically depending on the time elapsed since the occlusion of the coronary vessel. The parameters that changed over time were set to certain values according to table 6.1 at different stages shown in Fig. 2.6. A linear interpolation of the values in between allowed for a dynamic computation of the temporal changes. For the representation of hyperkalemia, the extracellular potassium concentration $[K^+]_o$ was increased. Acidosis was considered in the simulations by a reduction of the conductivities g_{Na} and g_{CaL} and a shift of the voltage dependence of the sodium channels $dV_{m,Na}$. The reduction of the intracellular ATP concentration $[ATP]_i$ and the increase of $[ADP]_i$ reflected the effects of hypoxia.

In this work, the formulation of $I_{K,ATP}$ was modified, since this current is inhibited by physiological concentrations of ATP [60]. For this purpose, the half maximum inhibition constant K_m was adjusted:

$$K_m = \left(-151.0919 + 75.5379 \cdot [ADP]_i^{0.256}\right) \cdot K_{m,factor} \qquad (6.1)$$

As in [148], the value of $K_{m,factor}$ varied in endocardial, M and epicardial myocytes reflecting the different sensitivities of these cell types to a lack of ATP and an increase of ADP [68]. The existing model was extended by adding parameter sets of the altered model parameters for phase 1b ($20 - 30\,min$ after occlusion) according to Pollard et al. [141]. Furthermore, the values of k_{NaCa}, P_{NaK}, V_{rel} and V_{maxup} were reduced in this later stage of ischemia [141]. Acidosis was responsible for the reduction of k_{NaCa}, whereas hypoxia led to decrease of P_{NaK}, V_{rel} and V_{maxup} after $20 - 30\,min$ of ischemia [142]. In this way, the impact of ischemia on ionic pumps and exchangers, as well as the intracellular calcium handling was considered. In contrast to [141], modifications of I_{bCa} and the nonselective calcium sensitive cation current $I_{ns,Ca}$ were not implemented, since the effects could be neglected and the latter current was not part of the ten Tusscher model, respectively. All parameter changes due to acute cardiac ischemia are summarized in table 6.1.

6.2 Modeling Ischemic Tissue

As already described in section 3.3, an anatomical model of the ventricles was derived from magnetic resonance images of a healthy volunteer in a previous work [163]. In order to model the effects of acute cardiac ischemia in the ventricles, the main coronary arteries were segmented. For this purpose, a seed point was set in the aorta, which had the same intensity in the MR image as the coronary arteries, since they were filled with blood. The region growing method, which iteratively looks for values with similar intensity within defined thresholds in a neighborhood, was then used for the segmentation of the LCx and the LAD (see also Fig. 6.2). However, the RCA could not be identified in the MR images due to the limited resolution. Based on these findings, models of heterogeneous ischemia effects could be created and occlusion of these vessels could be reproduced.

6.2.1 Heterogeneous Ischemic Tissue

In order to simulate transmurally varying ischemia effects, the ventricular wall was divided into 20% epicardial, 40% midmyocardial and 40% endocardial tissue, as presented in section 3.3. Consequently, the sensitivity of $I_{K,ATP}$ to changes of $[ATP]_i$ and $[ADP]_i$ was increased from endocardial towards epicardial tissue.

As presented in [148], the ischemic regions and the influence of the distance to the occluded coronary artery on the effects of ischemia were modeled using a so-called zone factor (ZF). It varied between 0 indicating healthy tissue in the normal zone (NZ) and 1, which represented the CIZ, where ischemia effects were most severe. A BZ was located between both zones with a gradual increase of the severity of ischemia effects towards the CIZ. Furthermore, the course of the different ischemia effects hypoxia, hyperkalemia and acidosis varied across the BZ (see Fig. 6.2). After only 10% of the BZ, the effects of hypoxia were fully present, whereas the effects of hyperkalemia linearly increased across the entire BZ. The effects of acidosis started at 50% of the BZ and then underwent a linear course towards the CIZ [148].

In case of phase 1b ischemia $20 - 30\,\mathrm{min}$ after the occlusion of a coronary artery, the tissue conductivity was linearly reduced to 12.5% of the normal value towards the CIZ [143]. In this way, cellular uncoupling due to reduced gap junctional conductance was modeled.

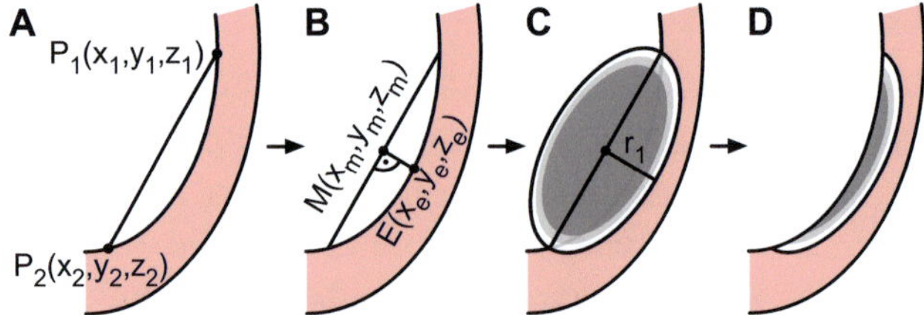

Figure 6.1. Creation of ellipsoidal ischemic regions in the concave ventricular wall. Two points P_1 and P_2 are set manually (A). On a plane perpendicular to the vector $\overrightarrow{P_1P_2}$, the point E on the endocardial surface with the shortest distance to the midpoint M of this vector is searched (B). The radius r_1 of the ellipsoid, which is placed afterwards, defines the transmural extent of the ischemic region. Iterative erosion of the ellipsoid leads to the incremental course of the ZF across the BZ (C). Only the overlapping parts of the ellipsoids with the ventricles remain (D).

6.2.2 Creation of Ellipsoidal Ischemic Regions

The ischemic regions modeled in this work were shaped as ellipsoids, so that the main axis of the ellipsoid reflected the course of the coronary artery that should be occluded. Furthermore, two radii defined the transmural extent and the size of the ischemic region.

For the creation of the ellipsoidal ischemic region, which had to be rotated in the coordinate system and aligned to the endocardial surface, a semi-automatic algorithm was implemented. Two points $P_1(x_1,y_1,z_1)$ and $P_2(x_2,y_2,z_2)$ were manually set at the endocardial surface defining the main axis $\overrightarrow{P_1P_2}$ of the ellipsoid (compare Fig. 6.1). The algorithm creating the ellipsoidal ischemic region required a concave ventricular wall between both points. The midpoint $M(x_m,y_m,z_m)$ of $\overrightarrow{P_1P_2}$ was computed defining the center of the ellipsoid. Then, a plane perpendicular to $\overrightarrow{P_1P_2}$ with $M(x_m,y_m,z_m)$ being on the plane was defined. The intersection point $E(x_e,y_e,z_e)$ of this plane with the endocardial surface, which had the shortest distance to $M(x_m,y_m,z_m)$, was determined afterwards. The vector $\overrightarrow{ME}$ defined the second axis of the ellipsoid and the radius r_1 determined the transmural extent. The radius r_2 defined the length of the third axis of the ellipsoid perpendicular to $\overrightarrow{P_1P_2}$ and $\overrightarrow{ME}$. An ellipsoid specified by the center and the three perpendicular axes was set in the ventricular model describing the maximal extent of the ischemic region. Afterwards, iterative erosion of the ellipsoid was used to create the gradual increase of the ZF towards the CIZ. The number of erosion steps defined the width of the BZ and also the increments of the ZF

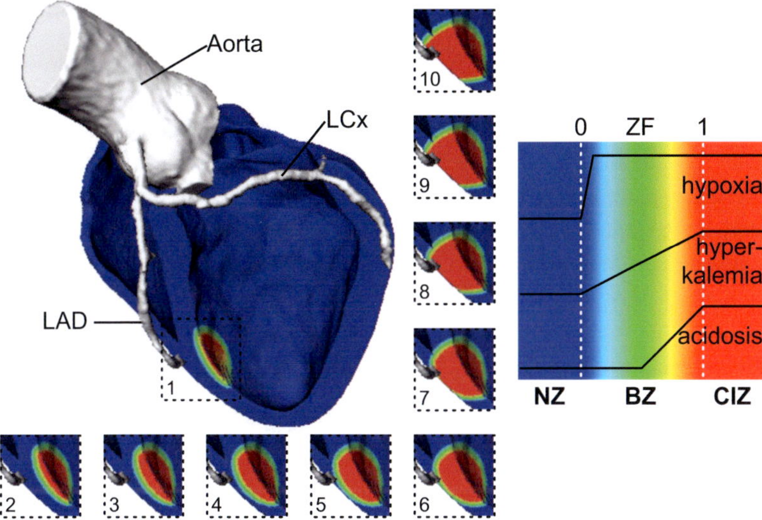

Figure 6.2. Ischemic zones in a model of the human ventricles describing an occlusion of the distal LAD. The transmural extent of the ischemic region was gradually increased from subendocardial (1) to completely transmural (10). The course of the ischemia effects hypoxia, hyperkalemia and acidosis varied across the BZ. Modified according to [17].

across the BZ. Finally, the parts of the ellipsoids, which did not overlap with the ventricular wall, were removed. In this way, ischemic regions describing the influence of the distance to the occlusion on the ischemia effects could be modeled.

In this work, the severity of the occlusion of the distal LAD was investigated. In Fig. 6.2, ten setups with varying transmural extent and radius r_1 of the ischemic regions, respectively, are presented. The radius r_2 was set to 12 mm in all cases. In the different setups, the endocardial surface of the ischemic regions was the same in all cases, only the volume of the affected tissue increased from subendocardial (1.36%) to transmural ischemia (3.83% of the volume of the left ventricle). The width of the BZ was set to 4 mm in all cases. The different stages of acute cardiac ischemia were initialized in a single-cell environment first. Afterwards, the excitation propagation in the ventricles was simulated during one cardiac cycle with a duration of 450 ms.

6.2.3 Forward Calculation of ECGs under the Influence of Ischemia

In case of phase 1a ischemia, the BSPMs and ECGs resulting from simulations of ischemia in the ventricles were forward calculated as described in section 3.3. However, the intracellular conductivities in the ischemic regions were reduced in case of phase 1b ischemia, as in the simulations of cardiac excitation propagation. For this purpose, high resolution tetrahedral torso models for each of the ischemia regions were created, since the intracellular conductivities in these regions differed from the healthy ones.

7

Modeling the Impact of Pharmacological Therapy on Cardiac Electrophysiology

The methods used for the investigation of the effects of three different pharmacological agents on cardiac electrophysiology are described in this chapter. The inhibition of certain ion channels due to the three compounds amiodarone, dronedarone and cisapride is delineated first. The effects of these drugs were investigated in multiscale simulations of control and electrically remodeled atrial, as well as in control and ischemic ventricular myocytes and tissue. Then, different markers used for the quantitative assessment of the antiarrhythmic potency of the drugs, as e.g. the vulnerable window, are presented. The corresponding results can be found in chapter 11.

7.1 Modeling Effects of Amiodarone

In section 2.5.2, an introduction to the effects of the class III multichannel blocker amiodarone was already given. The cardiac ion channels, which are affected by this drug, are summarized in Table 2.3. As noted in [87], the range of therapeutic concentrations of amiodarone is between 1 and 3 μM. Therefore, these two concentrations were used as low and high concentra-

Table 7.1. Reduction of ion channel conductivity (% of original value) due to antiarrhythmic agent amiodarone. Low concentration: 1 μM, high concentration: 3 μM.

conductivity (%)	setup 1		setup 2	
	low	*high*	*low*	*high*
g_{Na}	58.33	31.82	58.33	31.82
P_{NaK}	93.98	83.87	93.98	83.87
g_{Kr}	42.80	15.21	5.09	1.58
g_{Ks}	80.20	69.50	80.20	69.50
g_{CaL}	85.29	65.91	46.12	39.15
k_{NaCa}	76.74	52.38	76.74	52.38

tions for the simulation of the effects of amiodarone. With the help of the IC_{50} and nH values given in Table 2.3, the reduction of the maximal ion channel conductivities resulting from both concentrations could be computed using equation (2.3). Since the reported values of I_{Kr} and I_{Cal} varied in literature [79, 80, 82, 83], two setups representing different inhibition of the channels were generated. The resulting reductions of the maximal ion channel conductivities are summarized in Table 7.1. P_{NaK} and k_{NaCa} are the maximal NKA and NCX currents, respectively. Since amiodarone is recommended as a treatment option in case of myocardial ischemia [85] and AF [86], the influence on both pathologies was simulated in this work. For the simulation of the impact of amiodarone on atrial electrophysiology, the control and electrically remodeled Courtemanche model described in section 5.2 was adapted accordingly. The modified ten Tusscher model described in section 6.1 was adapted to simulate the drug effects in control and ischemic ventricular myocytes and tissue. The influence of amiodarone on ischemia was investigated only at 5 min after the onset of effects, since no excitation could be initiated in epicardial ischemic tissue after this stage. Ischemia effects were initialized in a single-cell environment first. In order to investigate the impact of amiodarone on the simulated body surface ECGs, only the control ventricular model was used, since the heterogeneous ischemia effects, which vary spatially and also temporally (compare chapter 10), would superimpose the drug effects.

7.2 Modeling Effects of Dronedarone

As described in section 2.5.3, dronedarone is also a multichannel blocker similar to amiodarone mainly used for the treatment of AF [104]. The ion currents, which are inhibited by dronedarone, are summarized in Table 2.4.

Table 7.2. Reduction of ion channel conductivity (% of original value) due to antiarrhythmic agent dronedarone. Low concentration: 150 nM, high concentration: 270 nM.

conductivity (%)	low	high
g_{Na}	93.12	80.37
g_{Kr}	32.19	22.88
g_{Ks}	86.31	82.38
g_{Kur}	86.96	78.74
g_{CaL}	99.10	95.61

As no information on the peak plasma concentration of dronedarone was available, the reported steady state plasma concentration between 150 and 270 nM [105] was used instead. With the low and high concentrations and the reported IC_{50} and nH values given in Table 2.4, the inhibitions of the maximal ion channel conductivities could be computed using equation (2.3). The resulting changes of the conductivities affected by dronedarone are listed in Table 7.2. These changes were integrated into the control and electrically remodeled Courtemanche model (compare section 5.2). In this way, the impact of dronedarone on cAF could be investigated in multiscale simulations ranging from the single cell up to 2D tissue patches.

7.3 Modeling Effects of Cisapride

Although cisapride was used as a gastroprokinetic agent (compare section 2.5.4), it also has an impact on cardiac I_{Kr} as shown in Table 2.4. The peak plasma concentration of cisapride ranged between 150 and 300 nM [110]. However, the effective concentration of cisapride inhibiting the ion channels was much lower, since more than 95% was bound to plasma proteins [107]. Consequently, the determined effective concentrations as well as the IC_{50} and nH values presented in Table 2.4 were used for the computation of the reduction of the maximal ion channel conductivities. Since cisapride prolonged the QT interval and caused torsades de pointes [109], its impact on ventricular electrophysiology was investigated in this work. For this purpose, the reductions of g_{Kr} were integrated into the modified ten Tusscher model described in section 6.1. As it exhibited similar QT prolongation as amiodarone, cisapride was compared to this antiarrhythmic drug.

Table 7.3. Reduction of ion channel conductivity (% of original value) due to gastroprokinetic agent cisapride. Low concentration: 150 nM, high concentration: 300 nM.

conductivity (%)	low	high
g_{Kr}	71.44	58.97

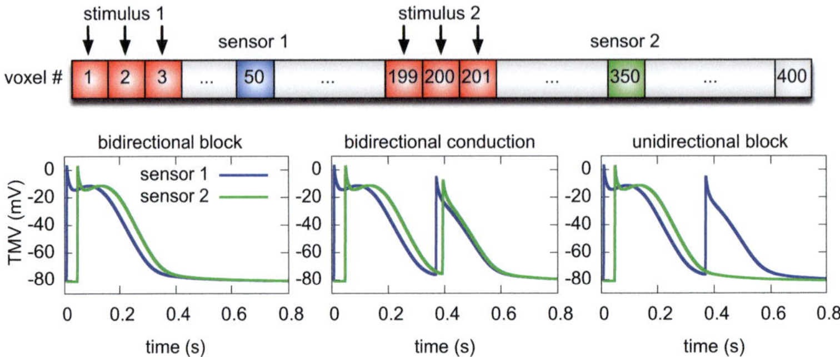

Figure 7.1. Detection of the vulnerable window, during which a unidirectional block can be initiated, using a 1D tissue strand. Depending on the stimulus time of the premature stimulus 2, three different effects could be observed: bidirectional block if the stimulus was applied too early, bidirectional conduction if the stimulus was too late, and unidirectional block during the vulnerable window.

7.4 Markers of Antiarrhythmic Potency of Pharmacological Agents

For the assessment of the antiarrhythmic potency of pharmacological agents, several markers, which try to quantify the ability of a compound to prevent or terminate rhythm disorders or to even promote them, exist. Some of the markers introduced in this section, as e.g. the vulnerable window (VW) or the WL, are of rather theoretical nature and are determined in computational studies mainly, whereas other markers are also used in clinical trials, as e.g. the QT interval.

7.4.1 Automatic Detection of the Vulnerable Window

Unidirectional blocks, which cause retrograde excitation propagation due to a premature stimulus at a wave back, are a necessary condition for reentrant circuits. The VW describes the spatio-temporal window, during which a uni-directional block can be initiated. The width of the VW indicates, how probable the initiation of a reentrant circuit is [183]. On the one hand, the site of the premature stimulus can be varied within a certain range and on the other hand, the time of stimulation causing a unidirectional block also varies within a specific interval. The VW can either be determined in a 2D tissue patch as e.g. in [184], in which reentrant circuits were initiated, or more simplistic in a 1D tissue strand, as introduced in [183].

In this study, the VW was determined in a 1D tissue strand consisting of $400 \times 1 \times 1$ voxels with a side length of 0.1 mm in case of atrial tissue and 0.2 mm in case of ventricular tissue, as shown in Fig. 7.1. At first, the models were initialized in a single-cell environment and then, four regular beats were initiated at the first three voxels of the 1D fiber (stimulus 1). Then, a premature stimulus was applied at the center of the tissue strand (stimulus 2), i.e. only the stimulus time was varied in this work, so that the VW described the time interval, during which unidirectional blocks could be initiated. Depending on the stimulus time, three different effects could be observed at two sensor sites. If the stimulus was applied too early, i.e. the tissue was completely refractory at the stimulus site, the premature stimulus was blocked in both retrograde and antegrade direction causing a bidirectional block. However, a second activation could be seen in both sensors, if the stimulus was applied too late, so that the stimulus site was completely excitable again (bidirectional conduction). In between both cases, the premature stimulus caused a retrograde excitation propagation only. The algorithm implemented in this work for the detection of the VW iteratively increased or decreased the stimulus time depending on the result of the previous premature stimulation. In this way, an upper limit of the stimulus time, after which bidirectional conduction occurred, and a lower limit, before which bidirectional block was observed, could be determined. The VW was defined as the interval in between, during which unidirectional blocks could be initiated.

7.4.2 Other Markers of Antiarrhythmic Potency

In addition to the VW, further markers, which indicate if a pharmacological agent has anti- or proarrhythmic properties, were investigated in this study.

Slope of the APD Restitution Curve

The slope of the APD or CV restitution is assumed to contain information on the probability of the initiation of fibrillation in tissue. This so-called restitution hypothesis states that steep slopes of the restitution curves (>1) cause oscillations of the WL. Due to this, wave breaks occur leading to chaotic excitation propagation patterns [185]. Consequently, flattening of the APD restitution is assumed to be antiarrhythmic [186]. In this study, the slope of the APD_{90} restitution was investigated at different diastolic intervals.

Wave Length

The wave length reflects the distance, which an excitation wave propagates during the refractory period, i.e. the WL determines the minimal distance between two consecutive wave fronts. Therefore, also the shortest possible path of a reentrant circuit is described by this measure of length. If the distance traveled by the excitation wave is smaller than the WL, the wave front collides with still refractory tissue and is eliminated afterwards. Since the WL influences the area occupied by a rotor, it is an indicator of the maintenance of rotors and gives an estimate of the maximal number of rotors in a defined area. Unphysiological changes of the WL promote the maintenance of reentrant circuits. The WL was calculated as the product of ERP and CV, which were determined as described in section 5.2, at different DIs.

Termination of AF

In case of simulations of cAF, the reaction of rotors to administration of drugs was tested in a 2D tissue patch. For this purpose, a stimulation protocol as described in section 5.4 was used to initiate four stable rotors in the electrically remodeled Courtemanche model. After the cross-field stimulus, two further stimuli at the wave backs of the reentrant circuits were applied to initiate additional rotors. Then, the resulting excitation propagation and the rotor trajectories (see section 5.5) were observed for 5 s. Furthermore, the dominant frequency was determined using pseudo-ECGs as described in section 5.6. This procedure was repeated, but after the initiation of the rotors, the effects of the pharmacological agents were included. In this way, the changes of the rotor trajectories, as well as the DFs could be compared between the different compounds and the cAF model alone.

QT Interval

The impact of pharmacological agents on ventricular electrophysiology could not only be investigated in the heart, but also on the body surface by forward calculation of ECGs, as described in section 3.3. Since some non-antiarrhythmic drugs were known to cause torsades de pointes, which was accompanied by a prolongation of the QT interval, this interval was introduced as a marker of cardiotoxicity [187]. The QT interval was determined as the time between the onset of the QRS complex and the end of the T wave. In this study, the onset of the QRS complex was defined by the endocardial stimulation profile, which initiated ventricular excitation propagation at $t = Q_{onset} = 0\,\mathrm{ms}$. Since the simulated ECG signals did not contain any measurement noise, a simple threshold voltage ($V_{thresh} = 0.01\,\mathrm{mV}$) could be used for the detection of T_{end}.

Part III

Results

8

Results: Parameter Adaptation of Electrophysiological Models

In this chapter, the results of the parameter adaptation of models of cardiac ion currents as described in chapter 4 are presented. At first, two methods for the solution of ODEs describing the gating of cardiac ion channels were investigated. Then, three different optimization algorithms for the minimization of the difference between simulated and measured ion currents were tested and compared alone as well as in combination. The ability of the algorithms to reconstruct known parameter values was assessed with synthetic ion current data and they were tested with different data of mutated ion channels. Finally, the sensitivity of the function to be minimized to changes of the adjustable parameters was analyzed.

8.1 Computing Times of Analytical and Numerical Solution of ODEs

The ODE (4.7) could be solved either analytically, since the transmembrane voltage was a piecewise constant function during the voltage clamp protocols used in this work, or numerically, which is the standard method in computational cardiology. The accuracy of the numerical approximation depends on the time step and error tolerances of the underlying ODE solver, which are specified in section 4.3. Therefore, only the computing times of both solution methods were compared in this work. They were measured with a 2.4 GHz Intel Xeon E5645 and 64 GB RAM under Mac OSX 10.7 (*Apple Inc., Cupertino, CA, USA*). The computing times of 1000 evaluations of the function to be minimized in equation (4.9) using the analytical or numerical solution were averaged and compared. One function evaluation took on average 0.35 ms in case of the analytical solution and 1.11 s in case of

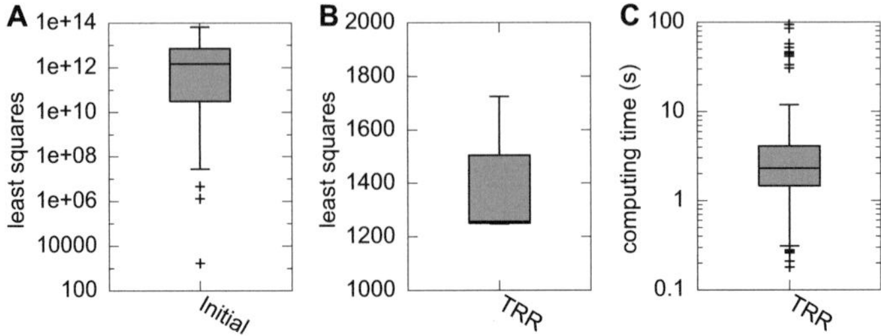

Figure 8.1. Box plots of the least squares before (A) and after the optimization using the TRR algorithm (B) resulting from 50 experiments with random initial parameter values. The corresponding computing times are shown in (C). The horizontal lines in the boxes, which surround the first and third quartiles, indicate the medians of the plotted data sets. The whiskers mark the most distant points within a range of 1.5 times the interquartile range. Plus signs indicate outliers.

the numerical approximation. As a consequence, only the analytical solution was used for the following optimizations due to the more than 3000 times smaller computing time of this method.

8.2 Test of Single Optimization Algorithms

Initially, each of the three optimization algorithms introduced in section 4.4 was investigated alone using the WT hERG measurement data described in section 4.1. Since the derivative-free PSO and GA, which were implemented in this work, had several parameters to be adapted, as e.g. the number of particles or iterations, different settings were tested first before comparing the algorithms.

8.2.1 Trust-Region-Reflective Algorithm

The results of 50 optimization experiments using the TRR algorithm with random initial parameter sets within the ranges defined in Table 8.1 are presented in Fig. 8.1. Before optimizing the parameters, the median least squares of the initial parameter sets was 1.497e12 and the outliers lay in a range from 1719.9 up to 6.636e13 (compare Fig. 8.1A). After the optimization using the TRR algorithm, the median least squares was reduced to 1256.6 and the variation of the resulting least squares was also reduced,

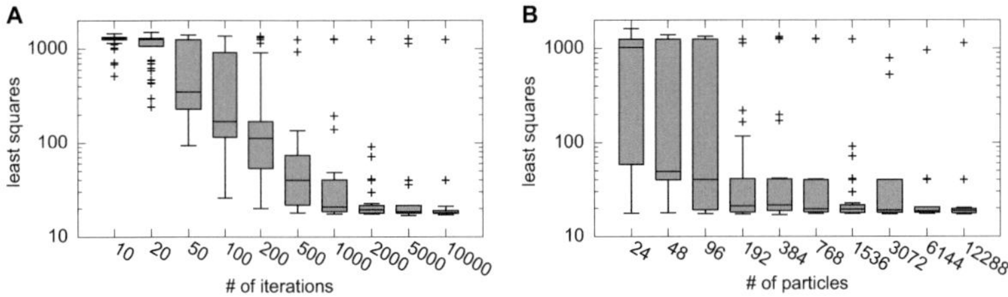

Figure 8.2. Least squares of the PSO algorithm with varying number of iterations (A) and particles (B) resulting from 50 experiments with random initial parameter values. The number of particles was set to $N = 1536$ in (A) and the number of iterations to 2000 in (B). Definition of box plots as in Fig. 8.1. Modified according to [12].

as shown in Fig. 8.1B. The computing time presented in Fig. 8.1C exhibited great variations depending on the initial parameter set with a median of 2.3 s. In all cases, the TRR algorithm stopped before the maximum number of iterations or function evaluations was reached, since the value of the function to be minimized finally changed by less than $1e - 11$, as defined in section 4.4.1. Therefore, the maximum number of iterations was sufficient allowing convergence of the optimization algorithm.

8.2.2 Adaptation of the Particle Swarm Optimization Algorithm

In case of the PSO algorithm, the number of iterations and particles N was varied in 50 experiments with random initial parameter values each to adapt the algorithm to the optimization problem (Fig. 8.2). The purpose of this adaptation was to find a combination, which on the one hand led to satisfactory adjustment of the parameters of the ion current model by minimizing the difference between simulated and measured current traces. On the other hand, a computing time as small as possible should be reached. In Fig. 8.2A, the number of iterations was increased from 10 to 10000, while the number of particles N was set to 1536. After 10 iterations, the median least squares was 1290.3, whereas it was reduced to 18.1 after 10000 iterations. The largest variance of the least squares could be observed around 100 iterations. Since the computer used for the analysis of computing times had 12 cores, the number of particles was increased in multiples of 24 in Fig. 8.2B to facilitate parallelizing the computation of the different particles per iteration. The number of iterations was set to 2000. Starting with 24 particles,

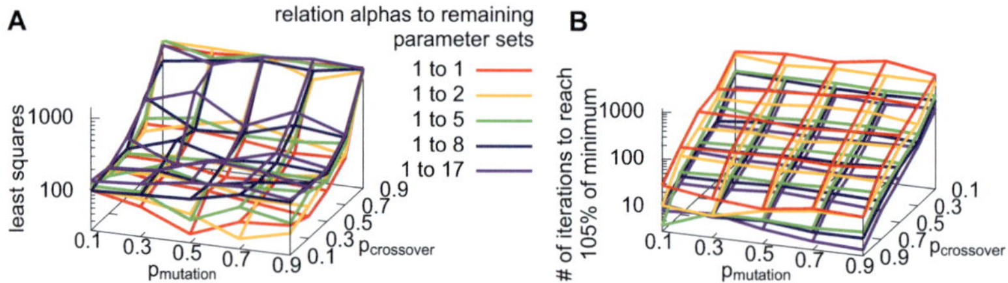

Figure 8.3. Median least squares (A) and median number of iterations to reach 105% of the minimum (B) of the GA resulting from 10 experiments with random initial parameter values. The probabilities $p_{mutation}$ and $p_{crossover}$, as well as the relation of alphas to remaining parameter sets were varied. The number of parameter sets was set to $N = 1026$ and the number of iterations to 2000. Modified according to [12].

a median least squares of 1030.3 with greatest variance was obtained. The median, as well as the variance, were reduced towards increasing number of particles. The median least squares was 18.7 in case of 12288 particles.

For the further use of the PSO algorithm, the number of particles was set to 1536 and the number of iterations to 2000, which was a compromise between the quality of the adaptation and the resulting computing time. A higher number of particles or iterations caused only a negligible improvement of the resulting least squares, whereas the computing time was increased by the factor of the increase of the number of particles and iterations.

8.2.3 Adaptation of the Genetic Algorithm

In case of the GA, the probabilities $p_{mutation}$ and $p_{crossover}$, as well as the relation of alphas to the number of remaining parameter sets were adjusted first. Then, the number of iterations and parameter sets N was determined. Fig. 8.3 presents the median least squares and median number of iterations to reach 105% of the current minimum resulting from 10 experiments with random initial parameter values with $N = 1026$ parameter sets and 2000 iterations. On the x and y axes, $p_{mutation}$ and $p_{crossover}$ were varied and the different grids correspond to different relations of alphas to remaining parameter sets. In most of the cases, the resulting median least squares shown in Fig. 8.3A increased with increasing $p_{crossover}$ and furthermore with decreasing relation of alphas to remaining parameter sets, i.e. if an alpha had the chance to pass its parameter values on to more remaining parameter sets. However, variation of $p_{mutation}$ caused no general trend of the resulting least squares. The

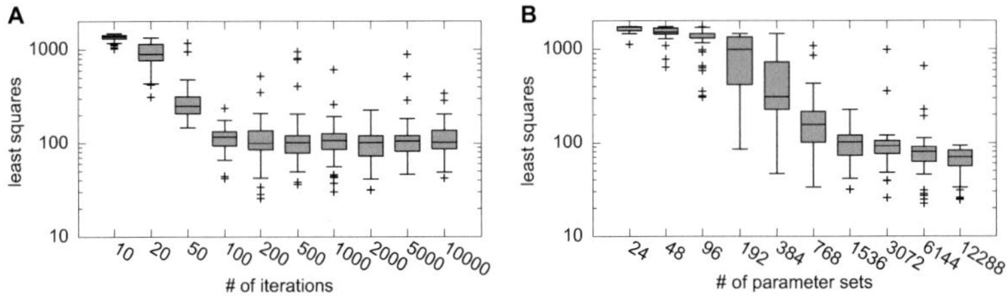

Figure 8.4. Least squares of the GA with varying number of iterations (A) and parameter sets (B) resulting from 50 experiments with random initial parameter values. The number of parameter sets was set to $N = 1536$ in (A) and the number of iterations to 2000 in (B). In case of 24 parameter sets, six outliers with values up to 1.44e7 were obtained, but the scale was adjusted for better visualization. Definition of box plots as in Fig. 8.1. Modified according to [12].

minimal median least squares of 39.4 was reached using $p_{crossover} = 0.1$ and $p_{mutation} = 0.5$ with a 1 to 1 relation between alphas and remaining parameter sets. The median number of iterations to reach 105% of the current minimum (shown in Fig. 8.3B) increased with decreasing $p_{crossover}$ and increasing relation of alphas to remaining parameter sets. However, variation of $p_{mutation}$ again depended on the other settings and presented no general trend. Using $p_{crossover} = p_{mutation} = 0.9$ with a relation of alphas to remaining parameter sets of 1 to 17, the convergence of the GA was fastest. In this case, 105% of the current minimum was reached after a median number of iterations of 4. As a compromise between the speed of convergence and the quality of the adaptation, a $p_{crossover}$ of 0.3, a $p_{mutation}$ of 0.5 and a 1 to 6 relation of alphas to remaining parameter sets were used. Compared to the setting causing the minimal median least squares, this configuration caused a 2.7 times higher least squares, but converged 7.6 times faster.

As in case of the PSO algorithm, the number of iterations and parameter sets N was varied in 50 experiments each with random initial parameter values to adapt the algorithm to the optimization problem (Fig. 8.4). In Fig. 8.4A, the number of iterations was increased from 10 causing a median least squares of 1368.6 to 10000 resulting in a median least squares of 102.4. The number of parameter sets N was fixed at a value of 1536. However, the median least squares remained in a similar range, if the number of iterations was increased starting from 200 iterations. In this case, the median least squares was already 101.2. Furthermore, the deviations from the median value also remained similar. The number of parameter sets N was varied in Fig. 8.4B from 24 causing a median least squares of 1721.9 to 12288 resulting in a median least squares of 69.9. The number of iterations was set to 2000. In case of 24 parameter sets, six outliers with values up to 1.44e7 were obtained, but the deviations from the median were reduced towards higher numbers of parameter sets.

On the one hand, the number of iterations was set to 500 in the following, since the GA showed faster convergence behavior than the PSO algorithm. On the other hand, the number of parameter sets N was set to 12288, which caused the best median least squares. The number of iterations was set to 500 to allow for convergence of the algorithm, since the number of parameter sets was higher and therefore the convergence behavior slower than in case of the 1536 parameter sets in Fig. 8.4A. As a consequence, this combination of number of iterations and parameter sets resulted in twice as many function evaluations as in case of the PSO algorithm.

8.3 Combination of Optimization Algorithms

In addition to the comparison of the three optimization algorithms alone, the combinations of the derivative-free PSO and GA with the TRR algorithm were evaluated using the WT hERG voltage clamp data, since the TRR algorithm alone was prone to end in local minima, as already mentioned in section 4.4.1. The algorithms were furthermore assessed regarding their ability to reconstruct synthetic data with known parameters. Finally, the algorithms were used to adapt the model parameters to two further voltage clamp current recordings of mutated ion channels. The twelve best parameter sets resulting from optimizations using the derivative-free algorithms were selected as an input for the subsequent TRR algorithm in case of combinations of the optimization algorithms.

8.3.1 Comparison of the Optimization Algorithms

Fig. 8.5 presents a direct comparison of the least squares, computing times and exemplary parameter values of g_{Kr} and Xr_{m1} resulting from 50 experiments with random initial parameter values using the different optimization algorithms alone and in combination. The median least squares (Fig. 8.1A) of the TRR algorithm alone was highest with 1256.6 compared to the GA

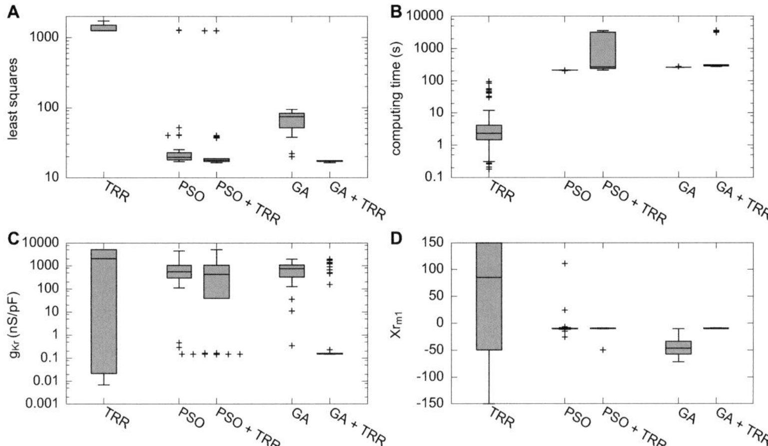

Figure 8.5. Comparison of the optimization algorithms alone and in combination. The least squares (A), computing times (B) and exemplary parameter values of g_{Kr} (C) and Xr_{m1} (D) resulted from 50 experiments with random initial parameter values. Definition of box plots as in Fig. 8.1. Modified according to [12].

with 73.7 and the PSO algorithm with 19.5. The deviation from the median value was also highest in case of the TRR algorithm, again followed by the GA. If the derivative-free algorithms were combined with the subsequent TRR algorithm, the median least squares and the deviations from this value were further reduced. The PSO combined with the TRR algorithm caused a median least squares of 17.7 and in addition, the smallest least squares reached by all algorithms (16.2) was obtained in three of the 50 experiments. However, the GA combined with the TRR algorithm showed almost no deviations around the median least squares of 17.5 being almost independent of the random initial parameter vectors.

The computing times of the different algorithms are depicted in Fig. 8.1B. In general, the TRR algorithm caused the greatest variance of the computing times. The computing times of the TRR varied depending on the initial parameter vector, whereas the derivative-free algorithms showed nearly constant computing times due to the fixed number of iterations. The TRR algorithm alone caused a median computing time of 2.3 s, the PSO algorithm needed 213.4 s and the GA 260.9 s. The evaluation of the function to be minimized in equation (4.9) contributed to on average 90.9% of the total computing time in case of the GA and even 96.2% in case of the PSO algorithm. The combination of the derivative-free algorithms with the TRR algorithm increased the median computing time of the PSO and TRR algorithm to 267.3 s and to 292.3 s in case of the GA and TRR algorithm.

The exemplary adjustable parameters g_{Kr} and Xr_{m1} are presented in Fig. 8.1C and D. The TRR algorithm alone caused various parameter values spanning nearly the entire range, which was defined for the parameter adaptation (compare Table 8.1). The median value of g_{Kr} was 2046.6 nS/pF in case of the TRR algorithm, whereas the PSO and GA algorithms found values of 551.7 nS/pF and 742.9 nS/pF, respectively. After use of the subsequent TRR algorithm, the median g_{Kr} values were reduced to 426.6 nS/pF in case of the PSO and TRR algorithm and to 0.149 nS/pF in case of the GA and TRR algorithm. However, the combination of the PSO and TRR algorithm increased the resulting variance, whereas the GA and TRR algorithm decreased the variance. The median value of Xr_{m1} (Fig. 8.1D) was 85.1 in case of the TRR algorithm and -46.2 in case of the GA, which presented also great variations of the values. In contrast, the PSO, PSO and TRR, as well as the GA and TRR algorithms caused similar median values of around -10 with almost no outliers.

The parameter values of the adaptation to the WT data using the PSO and TRR algorithm, which caused the smallest least squares, are summarized in Table 8.1. The corresponding current traces of the measured WT hERG data and the simulated I_{Kr} current can be seen in Fig. 8.7A. The simulated currents fit the measured curves well except for the fast upstroke of the current after the step to -120 mV. This part of the protocol could not be reproduced well by the Courtemanche model I_{Kr} formulation, which caused a slower upstroke.

8.3.2 Reconstruction of Synthetic Data

In order to assess the ability of the algorithms to reconstruct known parameters, synthetic voltage clamp data was generated as described in section 4.1. Then, each algorithm should reproduce the simulated synthetic data and adapt the parameters, so that the original parameter values were reconstructed. The resulting least squares of 50 experiments with random initial parameter values of the different algorithms are shown in Fig. 8.6A. In gen-

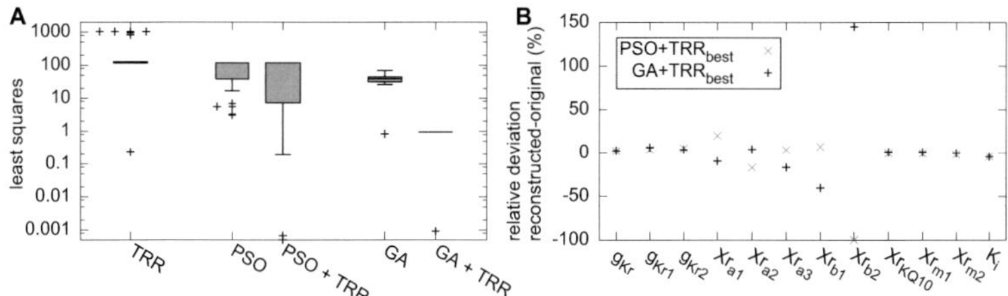

Figure 8.6. Reconstruction of simulated synthetic data set. Least squares of the different optimization algorithms resulting from 50 experiments with random initial parameter values (A). Definition of box plots as in Fig. 8.1. Relative deviation of reconstructed from original parameter values of the best optimizations using the PSO or GA combined with the TRR algorithm. Modified according to [12].

eral, qualitatively similar results as in case of the adaptation to the WT data could be observed. The TRR algorithm caused the greatest variations around the median value of 117.7. The PSO algorithm alone and combined with the TRR algorithm showed the same median value, however, the values of the outliers were reduced in case of the combination of the algorithms. The PSO combined with the TRR algorithm caused the two best least squares of all experiments (0.0005 and 0.0006). The smallest median least squares could be observed using the GA algorithm alone (39.6) and in combination with the TRR algorithm (0.95). Furthermore, the results of the GA combined with the TRR algorithm, which exhibited the third smallest least squares of 0.0009, was almost independent of the initial parameter vectors.

In Fig. 8.6B, the relative deviations between the original and reconstructed parameter values resulting from the best optimizations using the PSO and TRR, as well as the GA and TRR are presented. Most reconstructed parameters match the original values nearly perfectly except for the parameters $Xr_{a1}, Xr_{a2}, Xr_{a3}, Xr_{b1}$ and Xr_{b2}. These parameters, which influence the time constant τ_{xr} of the gating variable, showed larger deviations up to $+144.9\%$ and down to -99.8%. The parameter values giving the best reconstruction of the synthetic clamp data using the PSO and TRR algorithm are listed in Table 8.1. The corresponding current traces of the reconstructed simulation, which perfectly matches the original synthetic data, is depicted in Fig. 8.7B.

Table 8.1. Adjustable parameters of I_{Kr} with original values and their ranges. The parameter values presented in the last four columns are the result of 50 adaptations each with random initial parameter sets. Shown are only those resulting parameter sets, which caused the minimal least squares. The reconstructed parameters were obtained after adaptation to the synthetic clamp data. The WT, K897T and N588K parameter sets resulted from adaptation to the experimentally measured clamp recordings. Modified according to [12].

parameter	range		original	reconstructed	WT	K897T	N588K
g_{Kr}	0	5000	0.029411765	0.030068	0.1663	2615.9664	1427.7935
g_{Kr1}	−100	400	15	15.5909	11.2557	337.4943	323.2364
g_{Kr2}	−200	200	22.4	23.5206	21.8706	26.6261	28.1195
Xr_{m1}	−150	150	14.1	14.0285	−9.0557	−2.3033	−4.1286
Xr_{m2}	−150	150	6.5	6.4019	−11.9409	−22.8260	−21.6105
Xr_{a1}	0	2000	0.0003	0.00036	1970.4759	0.00003	0.00018
Xr_{a2}	−500	500	14.1	11.7296	260.1914	−19.1977	−35.1890
Xr_{a3}	−150	150	−5	−5.1760	7.5102	−8.5063	−9.7988
Xr_{b1}	−500	500	−3.3328	−3.5805	−23.0389	92.2019	78.3990
Xr_{b2}	−150	150	5.1237	0.0092	−16.4636	0.2711	0.000003
Xr_{KQ10}	−20	300	1.0	0.9976	−8.9526	16.0140	7.3604
$[K^+]_i$	0	3300	3000	2893.4584	35.9483	356.978	380.3644

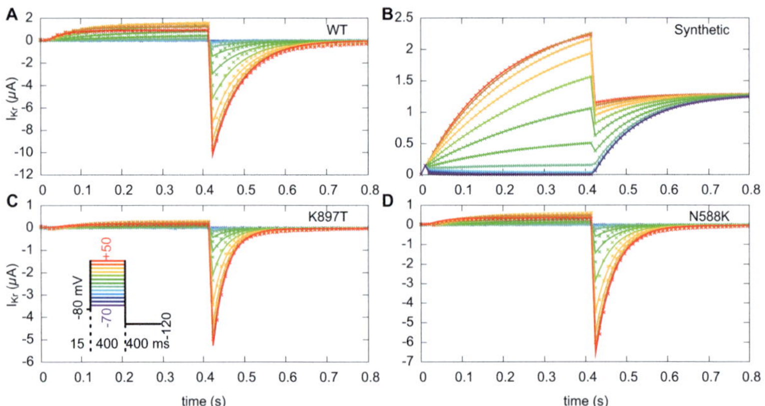

Figure 8.7. Measured hERG or synthetic I_{Kr} voltage clamp data ($\times$) and adapted simulated current traces (*solid lines*) with the corresponding clamp protocol. Voltage clamp recordings of WT (A), K897T (C) and N588K (D) mutations. Synthetic data (B) was generated by simulation of the same voltage clamp protocol using the original parameter values listed in Table 8.1.

8.3.3 Assessment of Algorithms Using Data of Mutated Ion Channels

The algorithms were furthermore tested with voltage clamp data of the K897T and N588K mutations of the hERG channel to investigate if the optimization algorithms and their settings were also suitable for the adaptation of other measurement data. In general, similar behavior of the different optimization algorithms as in case of the WT data could be observed. The resulting least squares were different, since the current amplitudes differed significantly. As in case of the WT data, the PSO combined with the TRR algorithm led to the minimal least squares, which was 1.82 for the K897T data and 4.28 for the N588K data. However, the GA combined with the TRR algorithm presented nearly constant least squares almost independent of the initial parameter vector with median values slightly above the best least squares. In Table 8.1, the adjustable parameters of I_{Kr} and their ranges, as well as the resulting parameter sets of the best adaptations are summarized. The corresponding current traces of the K897T and N588K mutations are shown in Fig. 8.7C and D. As in case of the WT data, the first part of the curve could be adjusted well, whereas the recovery from the second voltage step to $-120\,\mathrm{mV}$ was too fast for the model. In case of both mutations, this recovery was even faster than in the WT data. Nevertheless, the peak current amplitudes, which were reduced compared to the WT case, could be reproduced well.

8.4 Sensitivity Analysis

Finally, the local sensitivity of the function to be minimized in equation (4.9) was analyzed around the minimum of the adaptation of the model of I_{Kr} to WT voltage clamp data obtained by optimization using the PSO combined with the TRR algorithm. In this way, the sensitivity of the minimum to changes of one parameter at a time was investigated. The parameter values corresponding to the minimal least squares are given in Table 8.1. The function values and partial derivatives with respect to four exemplary parameters are presented in Fig. 8.8 to get an overview of how different the influences of the parameters on the minimum were. In Fig. 8.8A and B, the impact of variations of g_{Kr} on equation (4.9) could be observed. Since the function value depended quadratically on g_{Kr}, the resulting sensitivity was a linear function. The minimum, which was marked by $\times$ was close to zero and the function value increased quadratically towards higher values of g_{Kr}. In Fig. 8.8C and D, the influence of variations of the factor Xr_{a1} on the function to be minimized could be seen. Except for $Xr_{a1} = 0$, which caused extreme

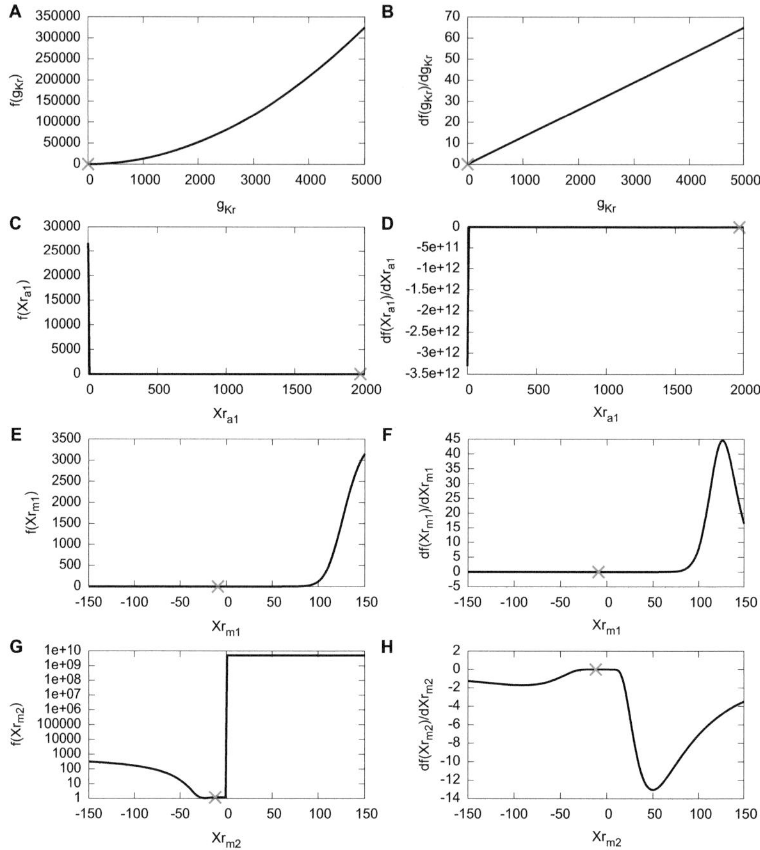

Figure 8.8. Sensitivity analysis of exemplary adjustable parameters. Shown are the function values and partial derivatives with respect to g_{Kr} (A, B), Xr_{a1} (C, D), Xr_{m1} (E, F) and Xr_{m2} (G, H) within the defined ranges specified in Table 8.1. The minimum, which was found by the optimization algorithm, was marked by $\times$.

changes and also a discontinuity of the function value, the sensitivity of this parameter was negligibly small, since the function value was nearly constant otherwise. The parameter Xr_{m1}, which was varied in Fig. 8.8E and F, showed also small sensitivity around the minimum. However, the function value increased towards the upper boundary of the parameter range causing a significant sensitivity there. The last exemplary parameter Xr_{m2} shown in Fig. 8.8G and H presented an almost constant function value immediately around the minimum resulting in a sensitivity of nearly zero. Towards more negative values, the function values slowly increased causing a small sensitivity there. However, a discontinuity could be observed at zero, since the values in equation (4.6) exponentially increased if Xr_{m2} became positive. Due to the discretization of the output of the sensitivity analysis, the expected peak of the sensitivity at zero could not be reproduced. Towards more positive values, a significant sensitivity was visible.

8.5 Discussion

In this study, the adaptation of parameters of a model of the rapid delayed rectifier current I_{Kr} to measured hERG and synthetic voltage clamp data was investigated. For this purpose, two methods for the solution of ODEs and three optimization algorithms were compared and evaluated regarding accuracy and computing time. Furthermore, the sensitivity of the detected minimum to variation of the parameter values was analyzed.

Since the transmembrane voltage was a piecewise constant function during the voltage clamp protocol used in this work, the ODEs describing the gating behavior of cardiac ion channels could be solved analytically. Compared to the numerical approximation, the computing time of this solution method was drastically decreased. In this way, use of complex population-based optimization algorithms, which are computationally expensive in general, was possible in an adequate amount of time. However, the computing times of the numerical and analytical solution of the ODEs also depend on the duration of the applied voltage clamp protocol, the number of samples and voltage steps. Due to this, the relative difference in computing times between both methods may vary depending on the data used.

The PSO and GA implemented in this work were adapted to the investigated optimization problem at first. In case of the PSO, the number of iterations and particles were varied for this purpose. The number of iterations was set to 2000 and the number of particles to 1536 as a trade-off between the quality of the adaptation and the computing time. In case of the GA, more settings were tested: the probabilities $p_{mutation}$ and $p_{crossover}$, the relation of alphas to remaining parameter sets and the number of iterations and parameter sets were varied. As a trade-off between speed of convergence and quality of the adaptation, $p_{mutation}$ was set to 0.5 and $p_{crossover}$ to 0.3 with a 1 to 6 relation of alphas to remaining parameter sets as in [172]. Compared to the PSO algorithm, the number of iterations could be decreased to 500 due to the faster convergence of the GA. However, the number of parameter sets was set to 12288 to obtain acceptable resulting least squares. Consequently, the product of number of parameter sets and iterations of the GA was two times higher than in case of the PSO. Nevertheless, the computing time of the GA was only 22% higher indicating that the parallelized computation of multiple parameter sets per iteration scaled differently than the serial calculation of iterations.

The results of the single optimization algorithms TRR, PSO and GA showed great differences depending on the initial parameter vectors. Due to this, the combination of the derivative-free PSO or GA with the TRR algorithm was tested using the WT hERG data first. In comparison to the derivative-free algorithms alone, the subsequent TRR algorithm generally improved the resulting least squares, but furthermore increased the computing time. As the median least squares was already small in case of the PSO algorithm alone, the improvement due to the TRR algorithm was relatively small. In contrast, the least squares resulting from the combination of the GA and TRR algorithm caused significant reduction of the median least squares. Additionally, the variance of the resulting least squares was smallest using the GA and TRR algorithm, i.e. the same minimum was found in almost all cases independent of the initial values. This effect can be explained by the high number of parameter sets covering great parts of the parameter ranges. Despite this, the PSO combined with the TRR algorithm found the best least squares in case of all data sets. The corresponding simulated current traces reproduced the measured WT current well, at least the first part of the voltage clamp protocol. The rapid recovery of the current after the voltage step to $-120\,\mathrm{mV}$ was too fast for the model of I_{K_r} of Courtemanche.

Both derivative-free algorithms successfully reconstructed most parameters of synthetic voltage clamp data with known parameter values. The PSO combined with the TRR algorithm caused the smallest least squares again. Only some parameters, which had an influence on the time constant τ_{xr}, showed greater deviations from the original values. Since these parameters depend on each other, they could not be uniquely identified using this voltage clamp protocol. Nevertheless, the resulting current traces of the reconstruction perfectly matched the original synthetic data. In case of the mutations, the results were qualitatively similar as in case of the WT data, i.e. the combination of PSO and TRR algorithm caused the smallest least squares, but the GA and TRR algorithm had a median least squares close to the smallest value with almost no deviations. The current traces were also reproduced well except for the fast recovery from the voltage step to $-120\,\mathrm{mV}$.

If the resulting parameter values of Table 8.1 are compared, some parameters as e.g. Xr_{a1} show large differences between the current recordings of WT and mutated ion channels. This discrepancy can be explained by the low sensitivity of this parameter over nearly the entire range used for the optimization. In general, some of the parameters presented low sensitivity or even discontinuities complicating the adaptation of the models of cardiac ion currents to measured voltage clamp recordings. Due to this, the combination of a derivative-free optimization algorithm for the selection of good initial parameter vectors for the subsequent minimum search using the TRR algorithm was better than the algorithms alone. Nevertheless, either a high number of particles have to be used to find the best parameter sets, or the best result of many random experiments as in case of the PSO combined with the TRR algorithm has to be considered.

In previous studies, different optimization algorithms were used to adapt the parameters of entire models of cardiac electrophysiology to reproduce AP or restitution curves of cardiac myocytes [188–191]. For this purpose, different optimization strategies with population-based or less complex algorithms were used. However, the amount of available information within the data used for the parametrization determines the maximal number of adjustable parameters that can be uniquely identified [167]. As a consequence, either the model that should be parametrized has to be simple with few adjustable parameters as in e.g. [190], or only some of the parameters as e.g. the maximal conductivities can be identified [188]. Derivative-free algorithms for the parameter adaptation of cardiac ion currents as the PSO algorithm were already used in [192], wheras a GA was investigated in [172]. In [2], the TRR algorithm was compared to other algorithms implemented in *Matlab*, whereas in [3], the influence of the analytic solution of the ODEs on the optimization using the TRR algorithm was investigated. In [4], the derivative-free GA and PSO were compared. However, this is the first comprehensive comparison of both derivative-free algorithms alone and in combination with the TRR algorithm used for the adaptation of cardiac ion currents to measured clamp recordings.

In this study, the I_{Kr} formulation of the Courtemanche model was used, although it is a Hodgkin-Huxley type model with only one activating gating variable. Due to this, the fast recovery from the voltage step to $-120\,\text{mV}$ could not be reproduced well by the model. However, the effects of mutations of cardiac ion channels should be investigated in multiscale simulations as described in section 9.3. Due to this, usage of the Courtemanche model, which is a well-established model applied for the simulation of e.g. effects of AF as shown in [13], was reasonable, since the K897T and N588K mutations analyzed in this work are known to favor AF [193, 194]. Despite this, a more sophisticated ion current model as e.g. a Markov model would probably improve the results of the adaptation to measured currents. Another limitation of this work is the use of hERG measurement data for the parametrization of a model of I_{Kr}, since hERG is only the pore-forming α-subunit of the channel. In addition, the oocyte voltage clamp measurements were performed at room temperature, which might probably not reflect the behavior of human I_{Kr} well.

Results: Modeling Atrial Fibrillation

The results of multiscale simulations of AF as described in chapter 5 and in [13, 120] are presented in this chapter. Initially, the general properties of five models of human atrial electrophysiology introduced in section 3.1.1 were benchmarked. Then, the ability of the different models to reproduce cAF-induced remodeling was investigated in multiscale simulations ranging from the single cell up to 2D tissue patches, in which a reentrant circuit was initiated. Finally, the effects of genetic defects of the ultra-rapid delayed rectifier potassium channel I_{Kur} were studied in multiscale simulations using the modified Courtemanche model of human atrial electrophysiology.

9.1 General Properties of Models of Atrial Electrophysiology

The five models of human atrial electrophysiology (Courtemanche, Nygren, Maleckar, Koivumäki and Grandi) were benchmarked regarding general model properties, which were also important for the simulation of AF.

9.1.1 Long Term Stability

Initially, the five models of human atrial electrophysiology were benchmarked regarding the numerical stability of model parameters over long simulation times. For this purpose, relative changes of AP parameters compared to published initial values were observed. Fig. 9.1A shows how the resting membrane voltage $V_{m,rest}$ of the different models reacted to sudden onset of pacing with a BCL of 1 s after a quiescent period of 20 min, during which no external stimuli were applied. In this way, the models could reach a nearly stable state of $V_{m,rest}$ first. The Courtemanche model presented almost no

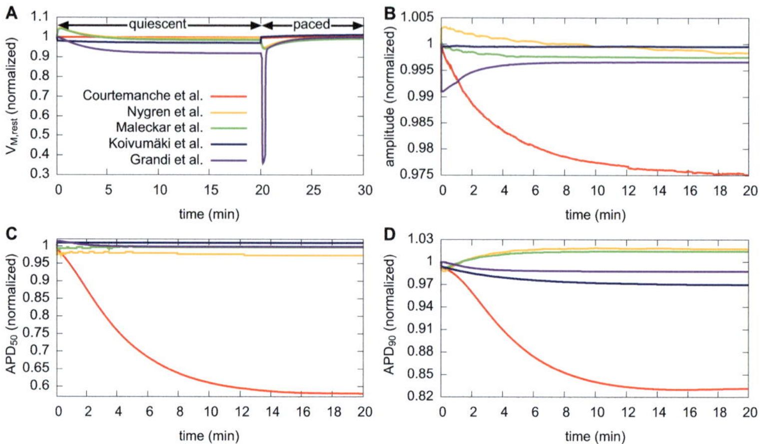

Figure 9.1. Long term stability of $V_{m,rest}$ (A), AP amplitude (B), APD$_{50}$ (C) and APD$_{90}$ (D) of models of human atrial electrophysiology. $V_{m,rest}$ was observed during 20 min without pacing and following 10 min of pacing at a BCL of 1 s. Amplitude, APD$_{50}$ and APD$_{90}$ were obtained during 20 min paced at a BCL of 1 s. All values were normalized to the first beat. Modified according to [13].

change during the first 20 min, whereas a slight decrease followed by an increase of $V_{m,rest}$ could be observed after the onset of pacing. The Nygren and Maleckar models showed an increase of $V_{m,rest}$ during the first minute and then a decrease. Then, the Nygren model presented a slight increase after 10 min, whereas the Maleckar model reached a stable $V_{m,rest}$ below the initial value. After the onset of external stimulation, $V_{m,rest}$ of the Nygren and Maleckar models decreased and then increased again. Within the first second, $V_{m,rest}$ of the Koivumäki model was reduced by nearly 1.2%. 10 min later, a stable $V_{m,rest}$ was reached. The first stimulus caused an elevation slightly above the initial value and then continuously increased until the end of the investigated period. $V_{m,rest}$ of the Grandi model gradually decreased to around 92% of the initial value after 10 min and remained stable until the onset of pacing. A rapid decline to 36% could be observed after the twelfth beat followed by a rapid and then a gradual increase up to the initial value 5 min later.

The AP amplitude, APD$_{50}$ and APD$_{90}$ were investigated during 20 min of pacing with a BCL of 1 s. Fig. 9.1B presents the amplitudes of the different models. The amplitude of the Courtemanche model gradually decreased to 97.5% of the initial value after 20 min. The Nygren and Grandi models showed an abrupt change of the amplitude after the first beat: the amplitude of the Nygren model increased by less than 0.5% and then gradually decreased below the initial value after 20 min, whereas the Grandi model presented a reduction of nearly 1% followed by a slow increase to 99.7% after 8 min. Then, the amplitude of the Grandi model remained constant until the end of the investigated period. The Maleckar model caused a slight reduction of the amplitude to 99.8% during the first 10 min, after which the results were stable. The amplitude of the Koivumäki model decreased marginally below the initial value during the first beats and remained constant.

The APD_{50} of the models over 20 min pacing with a BCL of 1 s can be found in Fig. 9.1C. The values of the Courtemanche and the Nygren model gradually decreased to 58% and 97.3%, respectively. The Maleckar, Koivumäki and Grandi models reached steady states close to the initial values after small adaptations of the APD_{50} during the first minutes.

In Fig. 9.1D, the APD_{90} of the five atrial models is presented during 20 min of stimulation with a BCL of 1 s. The APD_{90} of the Courtemanche model decreased by 17% during the first 16 min and then slightly increased until 20 min without reaching a steady state. The Nygren and Maleckar models caused a slight reduction of the APD_{90} during the first minute followed by an increase of approximately 2% until 10 min, after which the values remained stable. The Grandi model showed similar behavior with opposite sign: slight increase of the APD_{90} during the first beats followed by a reduction to 98.7% of the initial value. The APD_{90} of the Koivumäki model gradually decreased to 97% after 20 min.

9.1.2 AP and Calcium Transient Alternans

The phenomenon of alternans was investigated by stimulating the different models for 30 s at different BCLs (compare Fig. 9.2). The single cell APD_{50} restitution curves of the five atrial model are shown in Fig. 9.2A. The bifurcations of the traces of the Courtemanche, Koivumäki and Grandi models resulted from different numbers of stimuli, indicating an alternans of the APD depending on the number of stimuli. The Grandi model showed this bifurcation at a BCL of 0.5 s, whereas the Courtemanche and Koivumäki models

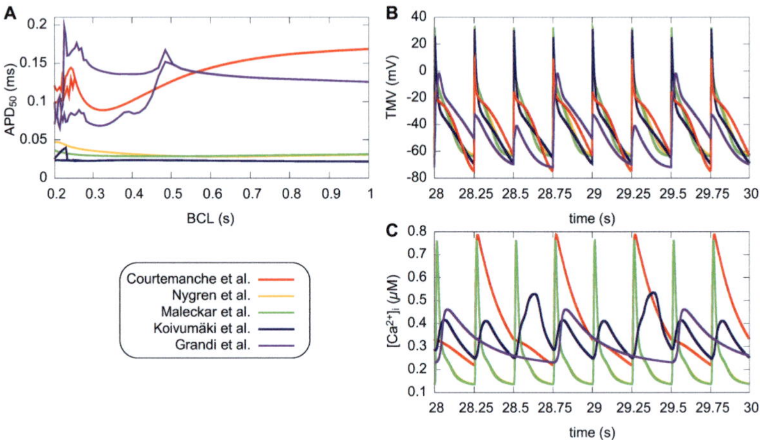

Figure 9.2. AP and calcium transient alternans of models of human atrial electrophysiology. Single cell APD$_{50}$ restitution curves over BCLs between 0.2 and 1 s (A). APs (B) and resulting calcium transients (C) over the last two seconds of 30 s pacing with a BCL of 0.25 s. Calcium transient curves of Nygren and Maleckar models overlap.

presented it at around 0.25 s. Consequently, the APs and calcium transients of the models were investigated during the last two seconds of 30 s pacing with a BCL of 0.25 s (see also Fig. 9.2B and C). A beat-to-beat alternans of the APs was observed in case of the Courtemanche model, as a peak of the calcium transient was initiated only every second beat. In this case, the intracellular calcium concentration [Ca^{2+}]$_i$ with a diastolic value of 0.2 μM and a peak value of 0.8 μM slowly decreased. The Nygren and Maleckar models did not show any alternans of the APs and the calcium transient, as already indicated by the APD$_{50}$ restitution curves. The calcium transients had a low diastolic value of 0.13 μM and a sharp peak with an amplitude of 0.75 μM. The Koivumäki model presented a reduced AP amplitude every third beat, which corresponded to an increased calcium transient peak (0.54 μM) in the previous beat. The diastolic [Ca^{2+}]$_i$ was slightly higher than that of the Courtemanche model. The most pronounced alternans could be observed using the Grandi model, which produced two APs with low amplitude and short duration after a longer beat. In this case, a peak of the calcium transient was initiated only every third beat with amplitudes in a similar range as the Koivumäki model.

Table 9.1. Computing times of the models of human atrial electrophysiology needed for the simulation of 1000 s with a BCL of 1 s without writing results to hard disk.

model	Courtemanche et al.	Nygren et al.	Maleckar et al.	Koivumäki et al.	Grandi et al.
time (s)	35.9	37.4	55.9	123.0	161.3

9.1.3 Computing Time

For the benchmark of models of atrial electrophysiology, the computing time plays a crucial role, since the simulation of AF on complex geometries can be a time-consuming process even with current computers. Table 9.1 summarizes the computing times of the five atrial models resulting from the single-cell simulations of 1000 s paced with a BCL of 1 s. To avoid delays during the simulation, no results were written to hard disk. The Courtemanche and Nygren models show similar computing times, as the number of underlying currents was the same. Compared to the Courtemanche model, which was the fastest, the computing time of the Maleckar model was increased 1.56-fold. The Koivumäki and Grandi models, which had a more complex description of the calcium handling, had 3.43 and 4.49 times longer computing times.

9.1.4 Discussion

Although the major goal of this work was to investigate which model of human atrial electrophysiology was suited best for the simulation of AF, an analysis of general properties of the models, as e.g. the long term stability of AP properties, was essential. However, these aspects were not analyzed in previous benchmarks of models of human atrial electrophysiology. In e.g. [131], only the ion concentrations were observed during 10 min of simulation. Depending on the requirements of the simulation, the AP properties should be stable over long periods for a robust usage, be able to reproduce the phenomenon of alternans to investigate its connection to rhythm disorders, or have small computing times for complex simulations using high-resolution geometries with a large number of elements.

The resting membrane voltage $V_{m,rest}$ of the Courtemanche, Grandi, Maleckar and Koivumäki model reached a steady state during 20 min without pacing, only the Nygren model showed a drift of $V_{m,rest}$. After the onset of stimulation, the Grandi model presented the most prominent changes of

$V_{m,rest}$ within the first minute. However, only the Koivumäki model showed a small drift of $V_{m,rest}$ after 10 min stimulation, the other models reached a steady state close to the initial value. The amplitudes of the Courtemanche and Nygren models still changed after 20 min pacing, whereas those of the Maleckar, Koivumäki and Grandi models reached a steady state close to the initial value. Only the Maleckar, Koivumäki and Grandi models presented a stable APD_{50} after 20 min pacing close to the initial value, whereas the Courtemanche model produced significant changes of the APD_{50} and did not result in a stable value after 20 min, which was also the case for the Nygren model. The APD_{90} of the Nygren, Maleckar and Grandi models reached a steady state during 20 min, in contrast to the Courtemanche and Koivumäki models. Consequently, only the Maleckar and Grandi models offered stability of all investigated AP parameters over longer periods and should therefore be used for long term simulations. However, slight modifications of the Courtemanche model published in 2009 [195] allow for long term stability of this model. Furthermore, other models were suited better for short term simulations and experiments, in which the model state was changed abruptly, since the Grandi model reacted sensitive to sudden onset of external stimulation after quiescence.

The Nygren and Maleckar models were not able to cause alternans in the single cell, whereas the Courtemanche, Koivumäki and Grandi models showed different alternans patterns of the APs and the calcium transients. Furthermore, the underlying mechanisms responsible for the alternans of the calcium transients differed, since the Courtemanche model has a phenomenological description of the SR calcium release depending on the transmembrane voltage. In contrast, the calcium handling of the Koivumäki and Grandi models is based on more recent findings enabling calcium-induced calcium release.

The aforementioned complex descriptions of the calcium handling of the Koivumäki and Grandi models cause a significantly higher computing time compared to the Courtemanche and Nygren models. Therefore, the Courtemanche and Nygren models were suited better for tissue simulations with a large number of elements.

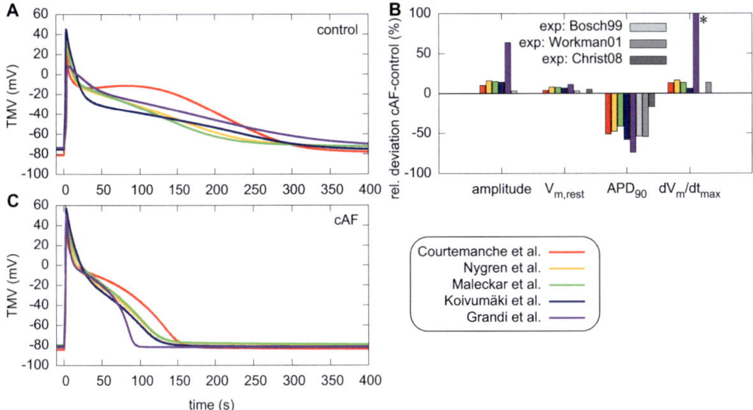

Figure 9.3. Single-cell APs of control and electrically remodeled myocytes. APs of control models of human atrial electrophysiology (A). Relative deviations between cAF and control AP parameters compared to experimental data of Bosch et al. [45], Workman et al. [196] (BCL $= 0.8$ s) and Christ et al. [197] (B). *: Relative change of dV_m/dt_{max} of Grandi model reached a peak of nearly 290%. APs resulting from model of cAF-induced remodeling (C).

9.2 Modeling Electrical Remodeling due to Chronic AF

The five models of human atrial electrophysiology were adapted as described in section 5.2 to reproduce effects of cAF-induced remodeling. The properties of the models of cAF were compared to the control models in multiscale simulations ranging from the single cell up to 2D tissue patches.

9.2.1 Single-Cell APs of Control and Electrically Remodeled Myocytes

The control APs of the five models of human atrial electrophysiology are presented in Fig. 9.3A. The AP of the Courtemanche model presented a pronounced spike-and-dome morphology, whereas the other four models produced triangular shaped APs. The Nygren and Maleckar models caused very similar APs. The AP of the Koivumäki model had a plateau phase with a significantly lower amplitude. The APs resulting from simulations using the Grandi model had a less pronounced overshoot than the other models and the myocyte repolarized in two phases causing a longer APD than the other models. In Fig. 9.3C, the APs of the cAF models are shown. All APs were shortened due to electrical remodeling and even the shape of the Courtemanche model was triangular. The relative changes of the AP parameters are visualized in Fig. 9.3B together with experimental values of Bosch et al. [45], Workman et al. [196] and Christ et al. [197]. The AP amplitude increased due to cAF. The relative changes of the amplitude of all models, especially the Grandi model, were above the experimental data. Furthermore, cAF-induced remodeling caused a shift of the resting membrane voltage $V_{m,rest}$ towards more negative values. The change of $V_{m,rest}$ of the Courtemanche model was in the range of experimentally observed values, whereas the other models were above the values. The APD_{90} was reduced due to cAF. The APs of the Koivumäki and Grandi models were shorter than the experimental data. In contrast, the APD_{90} of the Courtemanche, Nygren and Maleckar models were in the range of experimental values. As a consequence of cAF, the maximum upstroke velocity dV_m/dt_{max} was increased. Except for the Grandi model, all models lay in the range of experimental values. dV_m/dt_{max} of the Grandi model was increased by almost 290%.

The absolute values of the control and cAF AP parameters of the five atrial models are summarized in Table 9.2, which were compared to experimentally determined values of human atrial myocytes in Table 9.3. In the control case, the amplitudes of all models except for the Grandi model were in the range of experimental data. The amplitude of the Grandi model was approximately $30 - 40\,\text{mV}$ smaller than those of the other models. In case of cAF, only the Courtemanche model reproduced the experimentally determined amplitude. The other atrial models caused $10 - 15\,\text{mV}$ higher values. The resting membrane voltages $V_{m,rest}$ of the control and cAF Nygren, Maleckar, Koivumäki and Grandi models were similar to the experimental values. A more negative $V_{m,rest}$ could be observed using the control and cAF Courtemanche model. Compared to the Nygren, Maleckar and

Table 9.2. AP properties of control and electrically remodeled myocytes (BCL $= 1$ s).

AP parameter	Courtemanche		Nygren		Maleckar		Koivumäki		Grandi	
	control	*cAF*	*control*	*cAF*	*control*	*cAF*	*control*	*cAF*	*control*	*cAF*
amplitude (mV)	110.11	121.58	116.14	134.92	119.14	137.10	120.77	138.08	81.38	133.28
$V_{m,rest}$ (mV)	−81.0	−84.3	−74.2	−80.3	−73.8	−79.7	−76.1	−81.3	−73.5	−81.9
APD_{50} (ms)	165.16	74.99	29.55	30.14	29.77	35.33	20.98	29.17	125.30	39.54
APD_{90} (ms)	294.83	143.87	220.34	115.13	197.09	115.59	259.58	109.21	330.13	85.70
dV_m/dt_{max} (V/s)	186.58	211.38	149.77	174.96	160.66	182.47	168.54	179.26	92.50	359.53

Table 9.3. Experimentally determined control and cAF AP properties of human atrial myocytes. Values (where available) taken from Bosch et al. [45], Workman et al. [196] (BCL $= 0.8$ s) and Christ et al. [197].

AP parameter	Bosch et al.		Workman et al.		Christ et al.	
	control	*cAF*	*control*	*cAF*	*control*	*cAF*
amplitude (mV)	116 ± 3	120 ± 2				
$V_{m,rest}$ (mV)	-76.3 ± 2.2	-78.9 ± 2.9	-76.9 ± 2.1	≈ -77	-75.0 ± 0.4	-78.9 ± 1.1
APD$_{90}$ (ms)	255 ± 45	104 ± 9	209 ± 22	95 ± 12	≈ 344	287 ± 16
dV_m/dt_{max} (V/s)			203 ± 11	231 ± 16		

Koivumäki models, which exhibited similar APD$_{50}$, the Grandi and Courtemanche models presented much longer APD$_{50}$. The Courtemanche model had the longest APD$_{50}$ due to the spike-and-dome morphology of the AP, whereas the Koivumäki model showed the shortest duration caused by the low amplitude plateau phase. In case of the Grandi and Courtemanche models, the APD$_{50}$ was reduced due to cAF, but the values of the other models were increased. The APD$_{90}$ of the control Grandi model, which was longest among all models, best matched the measurements of Christ et al. [197], but the APD$_{90}$ was shortest in case of cAF reproducing the values of Workman et al. [196]. The APD$_{90}$ of the control Courtemanche model was closest to the measured values of Bosch et al. [45]. However, the Courtemanche model did not reproduce experimentally determined APD$_{90}$ in case of cAF. The control and cAF Koivumäki model caused APD$_{90}$ in the range of measurements of Bosch et al. [45], whereas the control and cAF Nygren and Maleckar models caused similar APD$_{90}$ reflecting the experimental values of Workman et al. [196]. The experimentally measured maximum upstroke velocity dV_m/dt_{max} [196] was reproduced best by the Courtemanche model, the values of the Nygren, Maleckar and Koivumäki models were significantly smaller. The Grandi model, however, showed a two times smaller dV_m/dt_{max} in the control case compared to the measured value, but had a more than 150 V/s faster upstroke than the other models in case of cAF.

9.2.2 Restitution Properties of Control and Electrically Remodeled Tissue

The restitution properties of the five models of human atrial electrophysiology were determined in simulations using the 1D tissue strand as described in section 5.2. The restitution curves of the APD$_{90}$, ERP, CV and WL of the

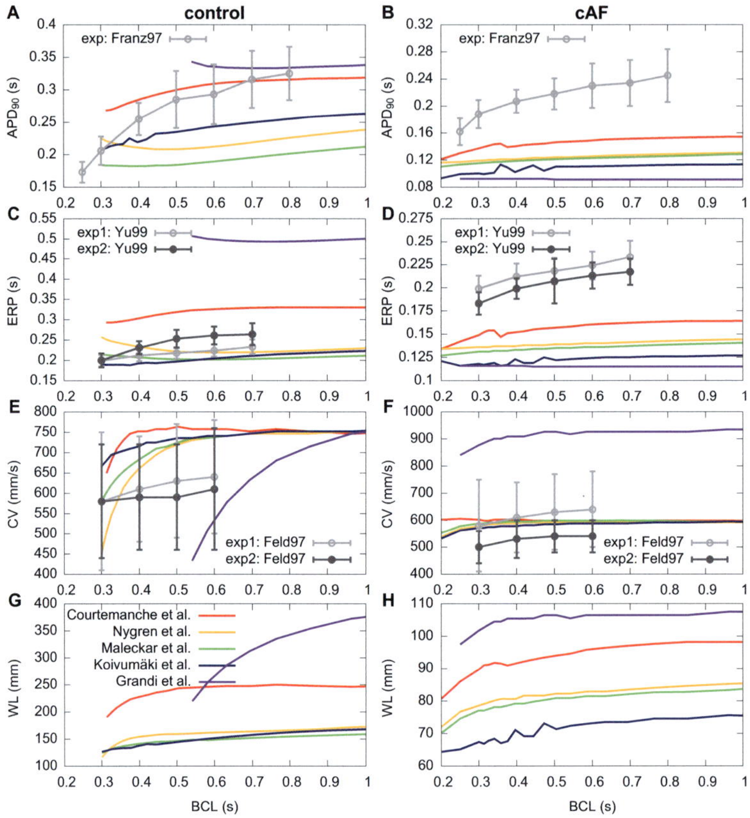

Figure 9.4. Restitution properties of control (*left*) and electrically remodeled (*right*) tissue simulated in a 1D fiber. APD_{90} restitution curves in comparison to experimental data of Franz [198] (A, B). ERP restitution curves compared to experimental data from the right atrial appendage (*exp1*) and the distal coronary sinus (*exp2*) of Yu et al. [199] (C, D). CV restitution curves compared to experimental data from the right atrial free wall (*exp1*) and the septum (*exp2*) of Feld et al. [200] (E, F). Restitution curves of the WL (G, H). Modified according to [13, 120].

control and cAF models compared to experimental data where available are shown in Fig. 9.4. In the control case, the Grandi model could not excite the tissue at every beat at BCLs of 0.54 s and lower. The Courtemanche model presented a limiting BCL of 0.31 s, whereas the other three models could be evaluated at BCLs of 0.3 s in the control case. In case of cAF, all models except for the Grandi model initiated APs down to BCLs of 0.2 s. The Grandi model was limited to a BCL of 0.25 s.

The APD_{90} restitution curves of the control and cAF models were compared to experimental data of Franz [198] (see Fig. 9.4A and B). The Courtemanche and Grandi models lay in the range of experimental values in the control case at high BCLs. However, the Nygren, Maleckar and Koivumäki models, which generally had a shorter APD_{90}, showed similar values as the experiments at a BCL of 0.3 s. The APD_{90} restitution curves of the Courtemanche and Koivumäki models had slightly flatter slopes than the experiments, whereas the Nygren, Maleckar and Grandi models showed an increase of the APD_{90} towards shorter BCLs. In case of cAF, the APD_{90} and also the slopes of the restitution curves of all models were smaller than in the experiments. The APD_{90} of the Grandi model was constant over the different BCLs.

In Fig. 9.4C and D, the ERPs of the atrial models are compared to experimental data of Yu et al. [199] originating from the right atrial appendage (*exp1*) and the distal coronary sinus (*exp2*). In the control case, the values of the Nygren, Maleckar and Koivumäki models were in the range of the experiments. Nevertheless, the ERP of the Nygren and Maleckar models increased towards shorter BCLs, whereas the Koivumäki model presented a decrease similar to the experimental data. The ERP of the Courtemanche model was around 50 ms longer than the experimentally measured values, but the slopes of the restitution curves were similar. The Grandi model produced an ERP, which was around 200 ms longer than the measured values and increased towards shorter BCLs. In case of cAF, the ERP of all models was more than 50 ms shorter than the experimental data. Again, the Courtemanche model reproduced the slopes of the measured restitution curves best. The slopes of the other models were significantly smaller.

Experimental data of Feld et al. [200] from the right atrial free wall (*exp1*) and the septum (*exp2*) was used for the comparison with simulated CV restitution curves of the control and cAF models (compare Fig. 9.4E and F). In the control case, the CV restitution curves of the Courtemanche, Nygren, Maleckar and Koivumäki models lay approximately within the standard deviation above the mean of the experimentally measured values, since the tissue conductivity of each model was adjusted to obtain a CV of around 750 mm/s at a BCL of 1 s. The measured CV slightly reduced towards short BCLs, whereas the simulated curves of the Nygren, Maleckar and Koivumäki models slightly decreased first and rapidly declined between BCLs of 0.4 and 0.3 s. The Courtemanche model presented a slight increase of the CV around a BCL of 0.5 s, which is called supernormal conduction [201], and also a rapid decline between 0.4 and 0.3 s. However, the CV of the Grandi model significantly decreased between BCLs of 1 and 0.54 s showing no stable CV values around a BCL of 1 s. In case of cAF-induced remodeling, the CVs of the Courtemanche, Nygren, Maleckar and Koivumäki models were in the range of experimental data, whereas the Grandi model caused more than 300 mm/s higher CVs than the other cAF models, which was higher than in the control case. The cAF Courtemanche model presented an increase of the CV towards short BCLs, whereas the other models reproduced the slopes of the measured restitution curves well.

As no experimental data of the WL was available, only the simulated values of the control and cAF models were compared to each other in Fig. 9.4G and H. The Grandi model caused the longest WLs followed by the Courtemanche model in the control and cAF cases. However, the control Grandi model caused a significant decrease of the WL between BCLs of 1 and 0.54 s, whereas the other models showed only a slight decrease of the WL towards short BCLs. The control Nygren, Maleckar and Koivumäki models caused similar WLs of around 150 mm, which was almost 100 mm shorter than the WL of the Courtemanche model and more than 200 mm less than the Grandi model at a BCL of 1 s. At a BCL of 0.4 s, which was the BCL used for the 2D rotor simulations, the Courtemanche model caused a WL of 227 mm, the Nygren model 154 mm, the Maleckar model 141 mm, the Koivumäki model 137 mm and the Grandi model did not initiate APs at every beat at this BCL. In case of cAF, the WLs of the Nygren and Maleckar models were similar and the Koivumäki model caused the shortest WL. All models presented a similar slope of the WL restitution curve in case of cAF. The WLs of the cAF models at a BCL of 0.4 s, which was used as pacing rate for the 2D rotor simulations, were: Courtemanche (92 mm), Nygren (81 mm), Maleckar (79 mm), Koivumäki (71 mm) and Grandi (105 mm).

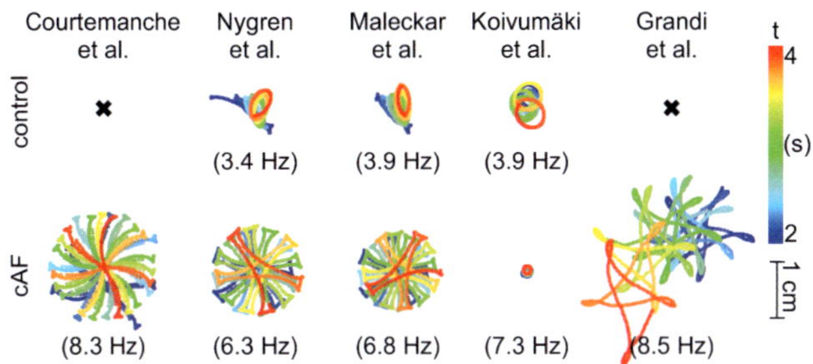

Figure 9.5. Rotor trajectories of the models of control and electrically remodeled tissue. The dominant frequencies derived from the pseudo-ECGs are specified below the trajectories in parentheses. Using the control Courtemanche and Grandi models, no reentrant circuit could be initiated. Modified according to [13].

9.2.3 Initiation of Rotors in Control and Remodeled 2D Tissue Patches

The possibility to initiate a single rotor in a 2D tissue patch using the control and cAF models of atrial electrophysiology was investigated as described in section 5.4. For this purpose, a cross-field (S1–S2) protocol should initiate a reentrant circuit. The control Courtemanche and Grandi models failed to produce a rotor in the 2D tissue patch due to their long WL. Using the control Nygren, Maleckar and Koivumäki models, as well as all cAF models, a reentrant circuit each could be initiated. In this case, the trajectories of the rotor centers were tracked as described in section 5.5. Furthermore, pseudo-ECGs were derived using two virtual electrodes and the corresponding DF was computed (compare section 5.6). The trajectories of the centers of all successfully initiated rotors are depicted in Fig. 9.5. The courses of the rotor centers were tracked between the second and fourth simulated second. The control Nygren, Maleckar and Koivumäki models caused elliptical rotor center trajectories with nearly the same size of less than 1 cm^2. The three models also caused similar DFs: the Nygren model presented the smallest value with 3.4 Hz and the two other models had a slightly increased DF of 3.9 Hz. In case of cAF, the Nygren and Maleckar models showed similar star-shaped trajectories occupying larger areas than in the control case. The DFs of both models were increased (+3.1 Hz). The cAF Koivumäki model produced a stable circular rotor trajectory, which was significantly smaller than in the control case. The DF of the Koivumäki model was increased (+3.4 Hz) in case of cAF. The Courtemanche and Grandi models caused curved star-shaped rotor center trajectories. The rotor center trajectory of the Grandi model occupied the largest area. Furthermore, the DF of the Grandi model was highest (8.3 Hz), but was only slightly higher than the DF of the Courtemanche model.

9.2.4 Discussion

In this study, the ability of the five models of human atrial electrophysiology to reproduce cAF-induced remodeling was investigated in multiscale simulations. For this purpose, the control and remodeled models were compared to each other and to experimental data. The resulting single cell APs of the different models were shortened due to cAF and showed a triangular shape. The AP of the Courtemanche model, which had a spike-and-dome morphology in the control case, was more triangular. The control and cAF AP amplitude of the Courtemanche model reproduced experimental data best, which showed an increase of amplitude due to cAF. The Nygren, Maleckar, Koivumäki and Grandi models closely matched control and cAF $V_{m,rest}$, however, the relative change due to cAF, i.e. a shift towards more negative values, was also reproduced by the Courtemanche model. Except for the Grandi model, the reduced APD_{90} of the electrically remodeled models reflected well the experimental values. The Grandi model appeared to be sensitive to changes of the amplitude of the external stimulus amplitude and $V_{m,rest}$, as the maximum upstroke velocity dV_m/dt_{max} was almost four times higher in case of cAF. Only the Courtemanche model closely matched the experimentally measured dV_m/dt_{max}.

The restitution properties of the models were investigated in a 1D tissue strand. The measured control APD_{90} restitution curves were reproduced best by the Courtemanche and Koivumäki model, although the slopes of the curves were too flat. In case of cAF, all models caused too short APD_{90} compared to the experiments. However, the slopes of the curves were similar to those of the experimental data. The ERP of the control Nygren, Maleckar and Koivumäki models were in the range of experimental values. Only the control and cAF Courtemanche model showed similar slopes of the ERP curves as compared to experiments. In case of cAF, the ERP of all models was below experimental values. All models except for the Grandi model lay in the range of experimentally measured CV, although the curves of the control models were too steep towards short BCLs. The cAF Grandi model caused a significantly increased CV compared to the control case, although the tissue conductivity was decreased as in the other models. Again, this was due to the lower $V_{m,rest}$, which caused an increased dV_m/dt_{max}. In all models, the WL was reduced due to cAF being shortest in the Koivumäki model and longest in the Grandi model, which was followed by the Courtemanche model. Consequently, no reentrant circuit could be initiated using the control Courtemanche and Grandi models, which was assumed to be more realistic than the ability of the control Nygren, Maleckar and Koivumäki models to initiate a rotor. The elliptical trajectories of the rotor centers and DFs were similar among the successful control models. The cAF models, which were all able to initiate a rotor, showed greater diversity regarding shape (from circular to star-shaped) and size of the rotor center trajectory and the DF.

Not all models considered in this work were published yet in previous benchmarks of models of human atrial electrophysiology. In [202], the Courtemanche and Nygren models have been compared regarding the difference in AP shape of the models, which was based on the different magnitude of the underlying currents, although the models relied on almost the same data. An analysis of the ionic currents was out of the scope of this work, which aimed at investigating the ability of the models to reproduce cAF-induced electrical remodeling. Cherry et al. [131, 203] investigated the differences of the Courtemanche and Nygren models regarding the ability to adapt to different pacing rates, coming to the same conclusion as in the present work that the APD of the Courtemanche model changed most with varying BCL (compare Fig. 9.4A). Furthermore, the initiation of rotors was also investigated in [203]. In case of electrical remodeling, stable reentrant circuits could be observed in both models. However, transient wave breaks could also be observed using the control Courtemanche model, which might be due to the use of tissue patches up to $30 \times 30\,$cm in contrast to $10 \times 10\,$cm in the present work. Furthermore, the CV used in the present work might have been higher resulting in a longer WL. These values were not stated in [203]. Therefore, no reentrant circuit could be initiated using the control Courtemanche model in this study in contrast to [203]. In [5], properties of the Courtemanche, Nygren, Maleckar and Koivumäki models were compared to each other. Furthermore, the minimal model published by Bueno-Orovio et al. [190], which was adapted by Weber [204] to reproduce the AP and restitution curve of the Courtemanche model, was also compared to the other models. However, the benchmark did not investigate rotor trajectories and DF in 2D simulations and did not include the Grandi model.

The benchmark of the models of human atrial electrophysiology showed that relatively small changes of the current formulations and variations of considered ion currents have a significant impact on the output of multiscale simulations. The results of the Courtemanche model were in good agreement with measured single-cell AP properties, as well as experimental data of the APD_{90} and CV. Nevertheless, the other models showed similar results as observed in some of the experiments, too, but none of the models could reproduce all experimentally investigated properties. The measured restitution properties were based on only one data set each. Therefore, more experimental data for the validation of the models of cAF-induced remodeling would be necessary. Depending on the purpose of the simulation, different models could be used. Regarding the initiation of rotors in a 2D tissue patch, the Courtemanche or Grandi model should be preferred, as only these cAF models were able to initiate rotors and not the control models, which is assumed to be the physiological case. Furthermore, the Courtemanche model was able to produce stable beat-to-beat alternans, which is linked to atrial rhythm disorders. In many simulation studies of AF, as e.g. [130–133], the Courtemanche model was already used. As already discussed in section 9.1, the computing time of the Courtemanche model was the shortest reducing costs of tissue simulations with large amount of elements. Besides, long term stability was not necessary during simulations shorter than a minute. Consequently, the Courtemanche model was used for the following simulations of effects of AF.

9.3 Modeling Genetic Defects Favoring AF

In this study, mutations of KCNA5 encoding the ultra-rapid delayed rectifier potassium channel introduced in section 5.3 were investigated in multiscale simulations. At first, the effects of the mutations were integrated into a mathematical description of the current at the ion channel level. Then, the resulting changes were investigated in the single cell, a 1D tissue strand and a 2D tissue patch.

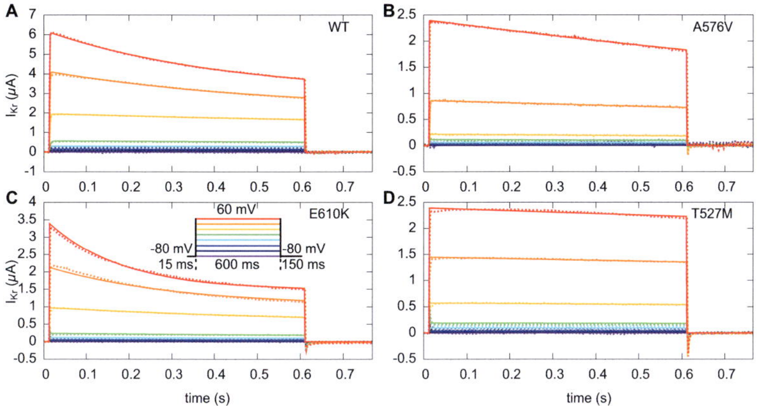

Figure 9.6. Patch clamp recordings (*dashed lines*) of WT and mutated I_{Kur} and adapted simulated current traces (*solid lines*) with the corresponding clamp protocol.

9.3.1 Adaptation of Model of Ultra-Rapid Delayed Rectifier Current to Mutations

The model of the ultra-rapid delayed rectifier current I_{Kur} of the Courtemanche model was adapted to the patch clamp data described in section 5.3 using the parameter adaptation framework introduced in chapter 4. The combination of the PSO and TRR algorithm was used to find the parameter values within the ranges defined in Table 5.2 resulting in the minimal difference between the simulated and measured current traces shown in Fig. 9.6. For the adaptation to data of WT and mutated channels, different ranges were defined: ua_{KQ10} and ui_{KQ10} were set to the values found after the adjustment to the WT data, since the same measurement conditions were assumed. Furthermore, the values of g_{Kur1} and g_{Kur2} of the adjustment to the WT data served as lower boundaries in case of the adaptation to data of mutated channels to prevent negative conductivities after the calculation of the relative changes due to the mutations using equations (5.13) and (5.13). All simulated current traces depicted in Fig. 9.6 matched the measured traces well (minimal least squares: 6.2 (WT), 1.9 (A576V), 2.8 (E610K) and 3.2 (T527M)). The WT currents shown in Fig. 9.6A presented the largest amplitudes and a slow exponential decrease during the voltage steps to 40 and 60 mV, respectively. However, the mutation A576V (compare Fig. 9.6B) caused nearly linearly

Table 9.4. Parameter values of the control model and resulting from adaptation to WT and mutated I_{Kur} and following calculation of relative changes due to mutations.

parameter	control	A576V	E610K	T527M
ua_{a1}	0.65	4.7228	198.9204	147.1922
ua_{a2}	10	-12.5983	107.4018	-28.0843
ua_{a3}	-8.5	-3.5609	-7.5221	-98.0457
ua_{a4}	-30	-2.4678	24.3074	127.6908
ua_{a5}	-59	-14.1038	-109.3183	-19.9413
ua_{b1}	2.5	2.5014	25.6124	21.3226
ua_{b2}	82	17.8828	34.7356	-23.3846
ua_{b3}	17	9.7316	74.7051	26.7549
ua_{m1}	30.3	-53.8125	-17.8970	-42.1249
ua_{m2}	-9.6	-9.6000	-9.6000	-9.6000
ua_{KQ10}	3	3	3	3
ui_{a1}	1	0.5408	0.9936	0.3239
ui_{a2}	21	1002.3652	-767.6070	1006.4243
ui_{a3}	-185	-184.9999	-185.0000	-185.0001
ui_{a4}	-28	-1066.5728	-17.4074	-85.0142
ui_{b1}	-158	-23.6979	-158.0000	-56.8848
ui_{b2}	-16	-0.3267	-21.9186	-0.2772
ui_{m1}	-99.45	-31.6312	-78.6915	-18.7487
ui_{m2}	27.48	0.2748	27.48	27.4799
ui_{KQ10}	3	3	3	3
g_{Kur1}	0.005	0.00504	0.00503	20.2543
g_{Kur2}	0.05	0.07845	0.0549	0.0500
g_{Kur3}	-15	-35.5380	-14.7700	-15.5589
g_{Kur4}	-13	-13.0469	-12.9319	-12.9121

decreasing currents with the maximal amplitude being more than 2.5 times smaller than the maximal WT current at the voltage step to 60 mV. However, this maximal current amplitude of the mutated channels was more than two times higher than that of the voltage step to 40 mV. In case of the mutation E610K shown in Fig. 9.6C, a fast exponential decrease could be observed at voltage steps to 40 and 60 mV. This part of the protocol could not be reproduced as well as the other parts of the protocol by the adapted simulations. Furthermore, the maximal current amplitude was reduced by less than two times compared to the WT currents. The T527M currents, which are depicted in Fig. 9.6D, were in a similar range than those of the A576V mutation. However, the currents resulting from the steps to 20 and 40 mV were increased compared to those of the A576V mutation. The currents decreased only slightly during the variable voltage steps.

Table 9.5. AP parameters of the control Courtemanche model and the model adapted to mutations of I_{Kur} (BCL $=$ 1 s).

AP parameter		control	A576V	E610K	T527M
amplitude	(mV)	105.6	105.4	105.0	104.3
$V_{m,rest}$	(mV)	-81.0	-80.4	-80.4	-80.4
APD_{50}	(ms)	172.4	225.5	224.8	225.7
APD_{90}	(ms)	298.4	290.9	293.1	293.9
dV_m/dt_{max}	(V/s)	190.9	197.7	198.2	198.2

The resulting parameter values describing the relative changes due to the mutations A576V, E610K and T527M are compared to the original control values of the Courtemanche model in Table 9.4. The temperature correction factors ua_{KQ10} and ui_{KQ10} had the same values as in the original model, since they were the same in the measurements of both WT and mutated channels. Further parameters as e.g. ua_{m2}, ui_{m2} or g_{Kur4} showed similar or the same values as in the control original model. In general, the parameter values of all mutations were in similar ranges, if they were close to the original model parameters. In case of some of the parameters as e.g. ui_{a4} or g_{Kur1}, very large differences between the mutations could be observed.

9.3.2 Effects of Ion Channel Mutations on Single Cell APs

The APs of the original control model and the models adapted to the I_{Kur} mutations A576V, E610K and T527M are presented in Fig. 9.7A and the AP parameters are summarized in Table 9.5. The APs of the myocytes with mutated channels looked similar in general. These APs furthermore exhibited an increased TMV after the overshoot causing a more pronounced spike-and-dome morphology compared to the control AP. This effect was strongest in case of the A576V mutation, whereas the two other mutations showed similar changes. The TMV returned to approximately 15 mV causing an elevated plateau phase and an APD_{50} prolonged by more than 50 ms. However, the repolarization phase was faster than in case of the control model, so that the APs of the mutated channels showed a slightly shortened APD_{90} compared to the control model. The upstroke velocities dV_m/dt_{max} were increased by around 4% in case of the mutations. Nevertheless, the AP amplitudes were similar.

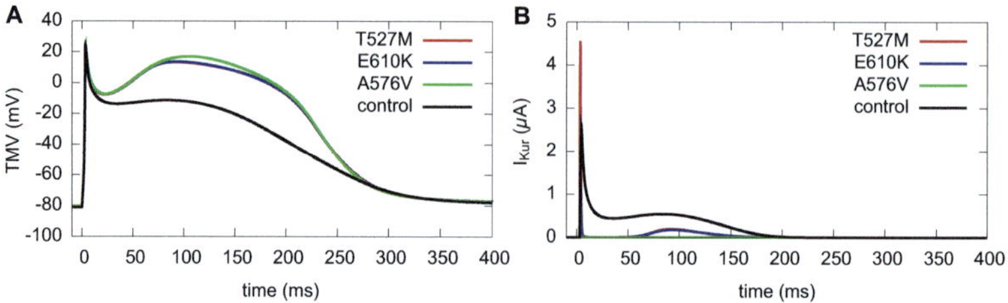

Figure 9.7. APs (A) and corresponding I_{Kur} current (B) resulting from mutations of KCNA5.

The underlying currents responsible for the differences are depicted in Fig. 9.7B. The course of the control I_{Kur} qualitatively followed the first part of the AP showing a fast upstroke and a short peak followed by a small dome with low amplitude going back to zero at around 200 ms. The mutations also presented a short current peak during the AP upstroke. The amplitude of this peak was increased in case of the T527M mutation, whereas the other mutations caused a reduction of the amplitude. During the plateau phase of the AP, the second rise of the currents of the mutated channels was significantly smaller in amplitude compared to the control current. After the initial peak, the current was nearly zero in case of the A576V mutation causing almost no rise of the current.

9.3.3 Restitution Properties of Tissue with Ion Channel Mutations

The restitution properties of the mutations of I_{Kur} presented in Fig. 9.8 were investigated in a 1D atrial tissue strand between BCLs of 0.3 to 1 s as described in section 5.2. In general, the mutations caused very similar restitution curves, which were hard to distinguish. Therefore, the changes of the restitution properties resulting from the mutations were described in general and not separately for each mutation. The mutations caused APs at every stimulus at BCLs down to 0.31 s, whereas the control model could be investigated at BCLs down to 0.3 s.

The APD_{90} restitution curve (Fig. 9.8A) of the control model started at around 0.32 s at a BCL of 1 s and slowly decreased to 0.25 and 0.27 s, respectively, at a BCL of 0.3 s. The APD restitution curve bifurcated at around 0.5 s due to negligible AP alternans, which slightly increased towards a BCL of 0.3 s. The APD_{90} of the mutations was around 20 ms shorter than that of

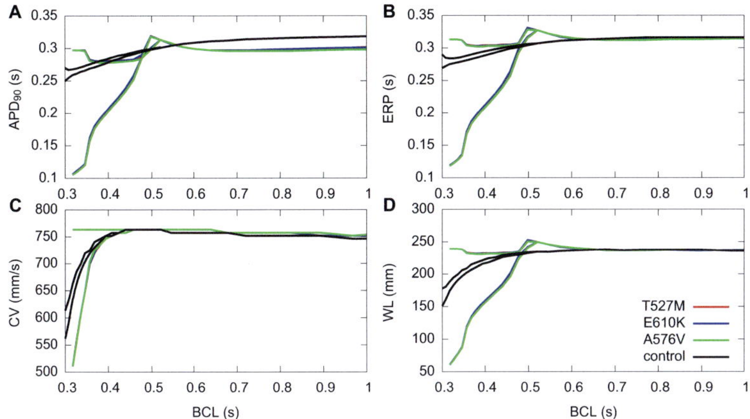

Figure 9.8. Restitution curves of APD$_{90}$ (A), ERP (B), CV (C) and WL (D) resulting from mutations of I_{Kur}. Bifurcations of the restitution curves reflect alternans at short BCLs.

the control model at a BCL of 1 s. The restitution curves slightly decreased towards a BCL of 0.65 s and then increased again. At a BCL of 0.52 s, the restitution curves bifurcated due to beat-to-beat alternans. Depending on the number of stimuli, the restitution curves either continuously decreased to approximately 0.1 s at a BCL of 0.31 s or increased to 0.32 s until a BCL of 0.5 s, afterwards decreased below the APD$_{90}$ of the control model and finally increased again to almost 0.3 s at a BCL of 0.31 s.

The ERP restitution curves shown in Fig. 9.8B looked qualitatively similar than those of the APD$_{90}$ restitution curves. However, the ERP of the control model was 0.315 s at a BCL of 1 s and decreased to 0.27 s or 0.29 s depending on the number of stimuli. The ERP restitution curves of the mutations were nearly the same as the control curve at high BCLs. However, the ERP started to increase at BCLs smaller than 0.65 s as in case of the APD$_{90}$ restitution curves of the mutations. Furthermore, the curves bifurcated in the same way at a BCL of 0.52 s, so that the ERP either decreased to 0.12 s or increased to 0.315 s again at short BCLs.

In Fig. 9.8C, the CV restitution of the control model and the mutations is depicted. The CV of the control model as well as the mutations started at 750 mm/s at a BCL of 1 s slightly increasing towards a BCL of 0.5 s. Then, the restitution curve of the control model bifurcated at a BCL of 0.45 s and the CV rapidly decreased to 560 or 620 mm/s at a BCL of 0.3 s. In case of the mutations, the restitution curves bifurcated already at a BCL of 0.47 s and either remained constant at around 760 mm/s or rapidly decreased to 510 mm/s.

The WL of the control model and the mutations was measured at various BCLs in Fig. 9.8D. All models presented a WL of approximately 240 mm at long BCLs. In case of the control model, the WL slightly decreased towards short BCLs starting at around 0.6 s. At 0.5 s, the control WL restitution curve bifurcated and showed small beat-to-beat alternans. The greatest deviations could be observed at a BCL of 0.3 s, where the WL slightly alternated between 150 mm and 175 mm. The mutations presented a prolongation of the WL up to 250 mm at BCLs shorter than 0.65 s. At a BCL of 0.52 s, the bifurcation of the WL due to alternans could be seen. The WL either decreased to 60 mm or shortly increased, decreased and then increased again to 240 mm. At a BCL of 0.4 s, which was used as the pacing frequency for the initiation of rotors in the following section, the WL of the control model varied between 215 and 221 mm, as well as between 155 and 230 mm in case of the mutations, depending on the number of stimuli.

9.3.4 Impact of Ion Channel Mutations on Initiation of Rotors

In addition to the previous analyses, the initiation of rotors as described in section 5.4 was investigated under the influence of mutations of the KCNA5 gene encoding the channels responsible for the ultra-rapid delayed rectifier potassium current I_{Kur}. For this purpose, a cross-field (S1-S2) protocol was used after the initialization of the models at a BCL of 0.4 s. Three regular line stimuli were applied to the tissue before the cross-field stimulus was placed. This stimulus was applied 50 ms after the end of the ERP of the previous regular excitation wave measured at the left border of the 2D tissue patch. In this way, the cross-field stimulus could initiate APs in parts of the tissue, which were already excitable again.

In Fig. 9.9A, the control Courtemanche model is shown. 345 ms after the initiation of the regular excitation wave, the cross-field stimulus was applied. In direction of the previous excitation wave, a conduction block due to the still refractory tissue could be observed. Therefore, the excitation could only propagate perpendicularly to the regular excitation wave. 100 ms later, the

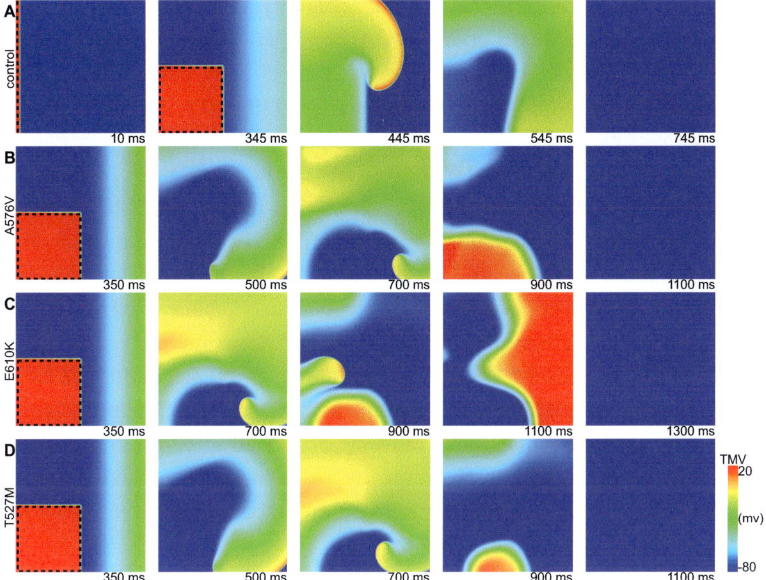

Figure 9.9. Initiation of rotors using the control (A), A576V (B), E610K (C) and T527M (D) models. For the initiation of a reentrant circuit, a cross-field (S1-S2) protocol was used. Stimulus sites are indicated by dashed lines. The TMV distributions of the three mutations and the control case shown in A–D were obtained at different times to fit the varying simulation times.

excitation had already entered the previously refractory tissue. However, the front of the excitation wave had propagated along the area of the cross-field stimulus, which was still refractory, and already left the tissue patch after 545 ms. Consequently, no reentrant circuit was initiated and the tissue was completely repolarized 200 ms later.

Since all mutations presented pronounced alternans at a BCL of 0.4 s as described in the previous section, the number of stimuli played a crucial role in the initiation of rotors. As in case of the control model, three regular line stimuli were applied first, so that the last regular beat had a long duration and fast CV. Due to this, the premature cross-field stimulus could cause an excitation wave with shorter WL making a reentrant circuit more probable. At a BCL of 0.4 s, the ERP of all mutations was 5 ms longer than in case of the control model, so that the cross-field stimulus was applied after 350 ms. After 500 ms, all mutations caused excitation waves entering the area of the cross-field stimulus, which was not refractory any more. The excitation waves showed more than a complete rotation after 700 ms. However, differences between the mutations could be observed already, being more clear at 900 ms. In case of the A576V mutation (Fig. 9.9B), the lower left part of the patch was excited by the rotor, but the previous excitation wave caused a conduction block, so that the wave front left the tissue patch, which was completely repolarized after 1100 ms. A similar pattern could be observed in case of the T527M mutation shown in Fig. 9.9D, where the conduction block was more pronounced. Due to this, a small part at the lower border of the patch and the upper left corner were still excited after 900 ms. Again, the reentrant circuit terminated and the tissue was completely repolarized after 1100 ms. In contrast, the E610K mutation showed no conduction block eliminating the excitation wave after 900 ms. Instead, a small new excitation wave could pass through a tunnel of refractory tissue at the upper left corner and the lower tissue border. However, only the back of the wave remained after 1100 ms causing no reentrant circuit any more, so that the tissue was completely repolarized after 1300 ms.

9.3.5 Discussion

In this study, the impact of the KCNA5 mutations A576V, E610K and T527M, which are associated with AF [176], on human atrial electrophysiology was investigated. The effects of the mutations were integrated into the I_{Kur} formulation of the Courtemanche model using patch clamp data of the mutated channels. The measured current traces could be reproduced well by the simulations with the adjusted parameters. Furthermore, only relative changes due to the mutations were integrated into the model of atrial electrophysiology to eliminate influences of the measurement conditions. The resulting increased value of g_{Kur1} in case of the T527M mutation was probably responsible for the increased peak current amplitude during the upstroke of the AP. Furthermore, some of the parameter values as e.g. ua_{m2}, ui_{m2} or g_{Kur2} were similar or equal to the control values in case of some of the mutations, which might be due to the defined parameter ranges. In case of e.g. g_{Kur2}, the lower boundary was determined by the value resulting from the adjustment to the WT data to prevent negative conductivity values. Due to this, the parameter value could not get smaller than the defined boundary, resulting in almost no relative difference compared to the original control parameter value.

The resulting APs under the influence of the mutations presented an elevated TMV during the plateau phase causing a more pronounced spike-and-dome morphology of the APs. Due to this, the APD_{50} was significantly increased, whereas the APD_{90} was slightly shortened due to the faster repolarization (stronger activation of I_{Kr} and I_{Ks}). In general, the underlying I_{Kur} was reduced in case of the mutations, only mutation T527M caused an increased current peak during the upstroke of the AP. All mutations presented nearly the same restitution properties. The main difference between the control model and the mutations was the pronounced alternans of the APD_{90}, ERP, CV and therefore also the WL. At BCLs shorter than 0.52 s, the restitution curves bifurcated due to beat-to-beat alternans. Consequently, the maximal difference in APD_{90} and ERP was almost 300 ms, the difference in CV 250 mm/s and in WL 180 mm, depending on the number of stimuli. Due to this, the WL of the mutations could be in a range similar to the control Nygren, Maleckar and Koivumäki models as shown in section 9.2.2, allowing the initiation of reentrant circuits. Using the same cross-field (S1-S2) stimulus protocol for the control model and the mutations, the control model failed to initiate a reentrant circuit, since the WL was too long compared to the dimensions of the patch. In contrast, a single rotor could be initiated in case of the mutations. However, the reentrant circuits were not stable over the time, since the mentioned alternans caused regions in the tissue patch with longer or shorter ERP depending on the previous diastolic interval. As a consequence, chaotic excitation propagation patterns could be observed, especially in case of the E610K mutation. In 3D tissue simulations, these chaotic patterns would probably evolve into stable fibrillation, since the atrial anatomy with obstacles and heterogeneous tissue properties would allow better pathways for the reentrant circuits than in case of a homogeneous 2D tissue patch with borders as in this study. Furthermore, the BCL, at which the patch was initialized could be reduced, so that the WL would be smaller favoring stable rotors. Although the reentrant circuits initiated in case of the mutations were not stable, the proarrhythmic properties mainly due to the alternans could be shown. Due to lack of electrophysiological measurement data of the mutations simulated in this work, the observed effects could not be compared to experimental data on the organ level.

In previous work, the heterozygous E375X mutation of KCNA5, which is also associated with AF [205], was modeled in [136] by a reduction of the maximal conductivity only. In the simulations, the mutation caused early afterdepolarizations at high pacing frequencies, which might induce AF. The effects of pharmacological inhibition of I_{Kur} on AF was investigated in measurements or using computational simulations in e.g. [132, 206]. Nevertheless, this is the first study, in which a biophysically detailed model reproducing the measured currents under the influence of KCNA5 mutations was used to investigate the effects of these mutations on multiscale simulations of human atrial electrophysiology.

In future work, the sensitivity of changes of the parameters on the function to be minimized in equation (4.9) could be analyzed. In this way, information on the origin of the observed large differences between some of the parameters (e.g. ui_{a4} or g_{Kur1}) of the different mutations could be gained. Perhaps, the adjustment of the biophysically detailed model of I_{Kur} with 24 adjustable parameters to the given patch clamp currents is an underdetermined system, i.e. the parameters could possibly not be uniquely identified. For this purpose, a different clamp protocol could provide complementary information. In addition to the 2D patch simulations, the initiation of AF could also be simulated using accurate anatomical 3D models of the human atria in future work to investigate the chaotic excitation patterns resulting from the mutations. The multiscale simulations of the effects of ion channel mutations presented in this study help to better understand the mechanisms responsible for the initiation of AF. Furthermore, the pharmacological treatment could be adapted to the mutations based on these simulations, allowing for patient-specific therapy of AF in the future.

10

Results: Modeling Effects of Acute Cardiac Ischemia

This chapter presents the results of multiscale simulations of the effects of different stages of acute cardiac ischemia in the ventricles as described in chapter 6 and in [16, 17]. At first, the influence of ischemia on single ventricular myocytes was investigated. Afterwards, the impact of the occlusion of the distal LAD on the ventricular excitation propagation was investigated at different ischemia stages and with varying transmural extent of the ischemic region. Finally, the corresponding forward calculated BSPMs and ECGs were analyzed, especially regarding shifts of the ST segment, which is an important diagnostic marker of ischemia and myocardial infarction [72].

10.1 Single-Cell APs at Different Ischemia Stages

The APs resulting from the parameter changes described in Table 6.1 are shown in Fig. 10.1A. Only the course of the transmembrane voltage (TMV) of epicardial myocytes is presented in this figure, since the changes due to ischemia are most prominent in this cell type. This was due to the higher sensitivity of $I_{K,ATP}$ compared to endocardial and M cells. Nevertheless, qualitatively similar but less pronounced effects could be observed in these cells.

Table 10.1. AP parameters of an epicardial myocyte at different stages of acute cardiac ischemia.

AP parameter	control	stage 1	stage 2	phase 1b
	0 min	*5 min*	*10 min*	*20–30 min*
amplitude (mV)	124.1	108.5	83.1	74.6
$V_{m,rest}$ (mV)	−85.6	−73.6	−64.3	−58.1
APD$_{50}$ (ms)	279.6	94.3	56.3	42.7
APD$_{90}$ (ms)	309.1	116.5	72.1	56.3
dV_m/dt_{max} (V/s)	94.7	61.4	39.4	37.2

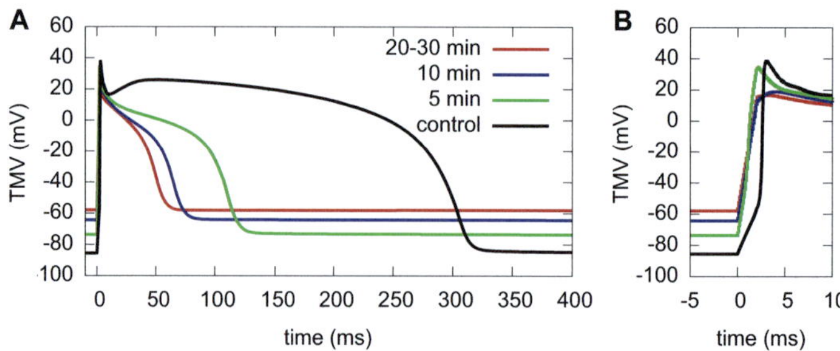

Figure 10.1. Action potentials (A) and upstroke phase of APs (B) of an epicardial myocyte at different stages of acute cardiac ischemia.

The main effects of acute cardiac ischemia on the AP, which are summarized in Table 10.1, were reduction of the APD, amplitude and maximum upstroke velocity dV_m/dt_{max}, as well as a rise of the resting membrane voltage $V_{m,rest}$. The slowed dV_m/dt_{max} due to ischemia is presented in detail in Fig. 10.1B. In the control myocyte, the TMV slowly rose until the threshold voltage of the sodium channels is reached. A fast upstroke of the TMV is visible afterwards. In case of ischemia, the membrane continuously depolarized with a slower upstroke velocity. In general, the impact of acute cardiac ischemia on the AP gradually increased with increasing time after the onset of the effects. However, the most significant changes of the APs shown in Fig. 10.1 could be observed between the control case and after 5 min of ischemia. In this case, the amplitude was reduced by 12.6%, the APD_{50} and APD_{90} were shortened by 66.3% and 62.3%, respectively. The dV_m/dt_{max} was slowed by 35.2% and $V_{m,rest}$ was shifted by 12 mV towards more positive values. In comparison, the changes during the following stage 2 after 10 min and phase 1b after $20 - 30$ min were smaller. Altogether, the amplitude was decreased to 60.1%, $V_{m,rest}$ was increased by 27.5 mV, the APD_{50} and APD_{90} were shortened by 84.7% and by 81.8%, respectively, and dV_m/dt_{max} was decreased by 60.7% after $20 - 30$ min of ischemia compared to the control case.

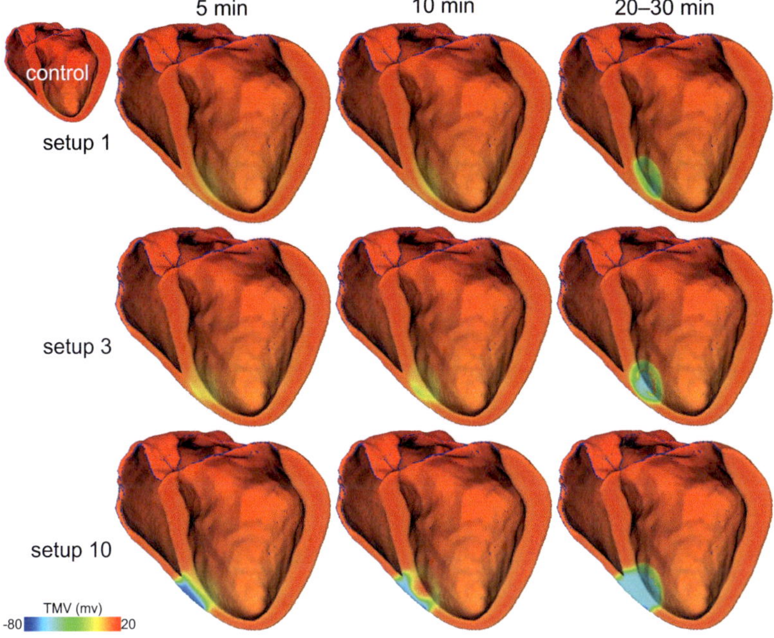

Figure 10.2. Ventricular TMV distributions ($t = 200$ ms of cardiac cycle) at different stages of ischemia (5 min, 10 min and $20 - 30$ min after occlusion of the distal LAD) with varying transmural extent of the ischemic region. Setup 1 was a subendocardial ischemia, setup 3 an intermediate ischemic region and setup 10 transmural ischemia. The corresponding ischemic regions are depicted in Fig. 6.2.

10.2 Modeling Ischemic Tissue

The impact of different stages of acute cardiac ischemia on ventricular excitation propagation was investigated using the ischemic regions depicted in Fig. 6.2. In this case, the severity of the occlusion of the distal LAD was analyzed by varying the transmural extent of the ischemic region ranging from subendocardial to completely transmural. The excitation propagation simulated in the ventricles with ischemic regions differed from the control case. The most prominent changes could be observed, if the entire ventricles were nearly homogeneously depolarized in the control case, which is illustrated in Fig. 10.2. There, distributions of the TMV of the control case and nine ischemia cases at $t = 200$ ms of the cardiac cycle are presented. In the three columns, the different stages (5 min, 10 min and $20 - 30$ min after occlusion of the distal LAD) of ischemia were grouped. The transmural extent of the ischemic region varied from subendocardial (setup 1) over an exemplary intermediate ischemic region (setup 3) to completely transmural (setup 10) in the different rows. During the first ten minutes of ischemia, the most significant changes were visible in case of transmural ischemia, since the epicardial layer was not activated there. However, setups 1 and 3 presented a slightly reduced TMV in the ischemic regions during phase 1a due to the lower amplitude and shorter duration of the APs. After 10 min of ischemia, a delayed activation of the endocardial layer resulting in a higher TMV than after 5 min of ischemia could be discovered in setup 10. During phase 1b, the ischemic regions could easily be identified, since the TMV was drastically reduced compared to the healthy tissue. In case of setup 1, only short APs with low amplitude were responsible for this effect, whereas complete conduction block occurred in the CIZ in case of setup 3 and setup 10.

10.3 Forward Calculation of ECGs under the Influence of Ischemia

The BSPMs shown in Fig. 10.2 resulting from the forward calculation of the TMV distributions during the ST segment can be found in Fig. 10.3. The torsos corresponding to different stages of ischemia and transmural extents of the ischemic regions were arranged in the same way as the corresponding TMV distributions in Fig. 10.2. In the control case, a nearly homogeneous BSPM of around 0 mV could be observed due to the completely depolarized ventricles depicted in Fig. 10.2 (inset). In general, Wilson chest leads V_3 and V_4 were closest to the ischemic region. Therefore, the most prominent changes of the body surface potentials could be found in this area. After 5 min of ischemia, only the transmural ischemic region (setup 10) caused a significant change of the BSPM compared to the control case. In setup 10, an elevation of the body surface potentials around the chest leads could be detected during all stages of ischemia, although the area of elevated potentials was smaller during phase 1b. In contrast, subendocardial ischemia (setup 1) resulted in a slight depression of the body surface potentials around the chest leads after 10 min becoming even more pronounced after $20 - 30$ min of ischemia. The BSPMs of the intermediate ischemic region (setup 3) did not

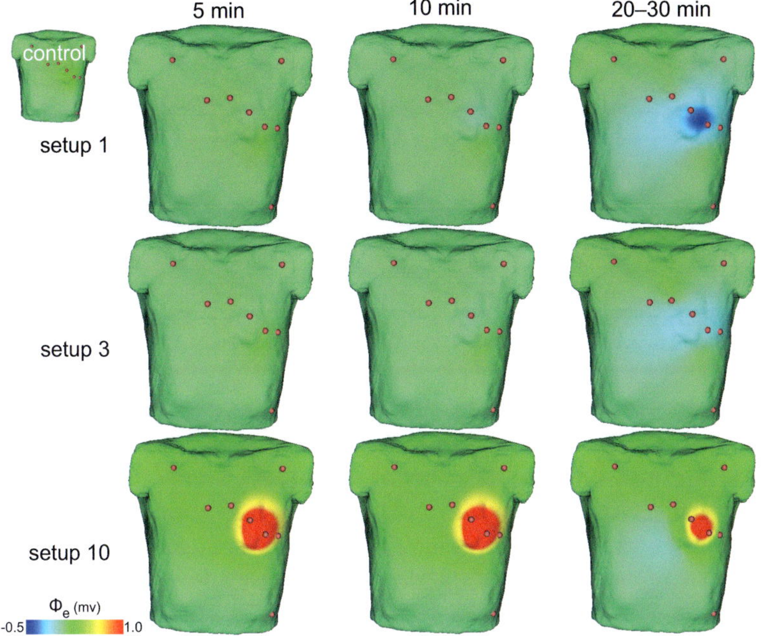

Figure 10.3. BSPMs ($t = 200$ ms of cardiac cycle) resulting from ventricular excitation propagation shown in Fig. 10.2. Shown are different stages of ischemia (5 min, 10 min and $20 - 30$ min after occlusion of the distal LAD) and varying transmural extent of the ischemic region.

differ from the control case till $20 - 30$ min after the occlusion of the LAD. After this time, a slight depression of the body surface potentials around the chest leads could be seen.

As the most prominent changes of the BSPMs due to the occlusion of the distal LAD were determined in Wilson chest lead V_4, the resulting ECGs of all simulated ischemia stages and transmural extents of the ischemic region are presented in Fig. 10.4. This figure gives an overview of the various changes of the ECG due to acute cardiac ischemia, mainly shifts of the ST segment, which are used as a diagnostic marker of ischemia and infarction. In addition, the TQ segment was slightly elevated due to the increased $V_{m,rest}$ in the ischemic regions. Moreover, the amplitudes of the QRS complexes and T waves, as well as the onset of the T waves were altered. In the control case, the ST segment was nearly a zero baseline, whereas subendocardial

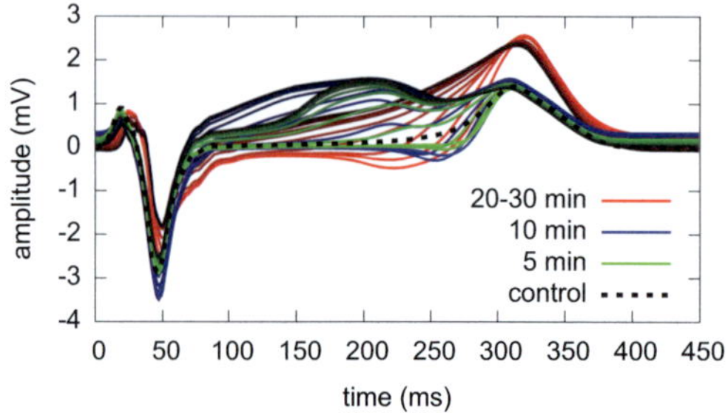

Figure 10.4. ECGs in the Wilson chest lead V_4 showing ST segment shifts resulting from different stages of ischemia (5 min, 10 min and $20-30$ min after occlusion of the distal LAD) and varying transmural extent of the ischemic region (*light tone:* subendocardial, *dark tone:* transmural).

ischemia (*light tone*) caused a slight depression of the ST segment. This effect was more pronounced at later stages of ischemia. However, transmural ischemia caused a pronounced elevation of the ST segment being most prominent after 10 min and smallest after $20-30$ min of ischemia. In between these shifts of the ST segment, some ischemia setups also caused a nearly zero baseline as in the control case. The 12-lead ECGs of these so-called "electrically silent" [1] ischemia cases (setup 3 at different ischemia stages) are presented in Fig. 10.5. The ECGs of the first 10 min of setup 3 are hard to distinguish from the control case. Only at the onset and end of the T wave, small differences could be noticed. However, the differences between the control case and setup 3 during phase 1b were more pronounced. There, even the ST segment was depressed in leads V_3 to V_5 in addition to changes of the QRS complex and the T wave.

10.4 Discussion

In this study, the influence of different stages of acute cardiac ischemia on a single myocyte, the ventricular excitation propagation and the forward calculated ECGs was investigated. The effects of acute ischemia on cardiac electrophysiology worsened in the course of the first 30 min. In the affected ventricular myocytes, the APs were shortened, the amplitude and maximum upstroke velocity reduced and the resting membrane voltage increased. In the ventricles, the transmural extent of the ischemic region and

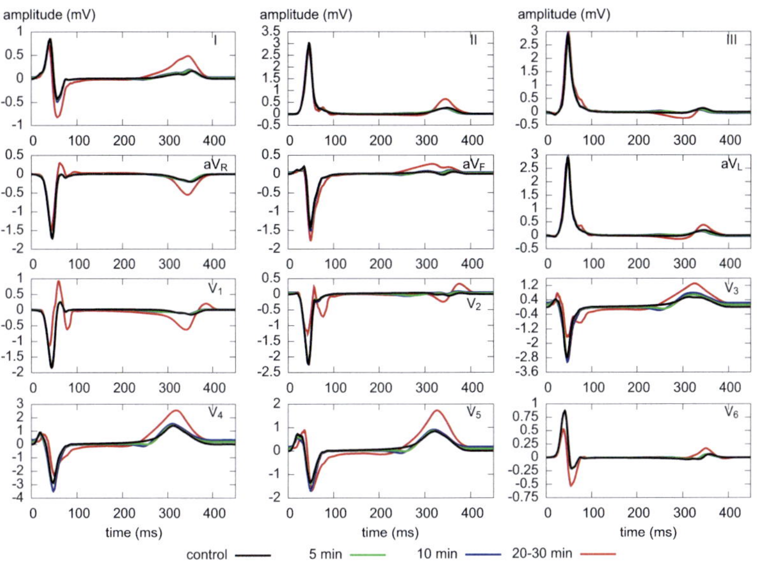

Figure 10.5. 12-lead ECG of control case and setup 3 (intermediate ischemic region) at different stages of ischemia. These ischemia cases were partly electrically silent, i.e. hard to distinguish from the control case.

the ischemia stage were varied. Subendocardial ischemia caused shorter APs with reduced amplitude in the endocardial layer with increasing effects towards phase 1b. The resulting injury currents were directed from the healthy epicardium towards the endocardium causing slight ST segment depression in leads close to the ischemic region. In case of transmural ischemia, the injury currents flowed from the less affected endocardium towards the more injured epicardium causing pronounced ST segment elevation. Cellular uncoupling during phase 1b effected a complete conduction block across the entire ventricular wall resulting in less pronounced ST segment elevation. In intermediate ischemic regions, injury currents even compensated each other producing similar ECGs as the control case.

In previous works, effects of phase 1a (e.g. [148]) or phase 1b (e.g. [141–143, 145]) on cardiac electrophysiology were investigated. However, the effects of different stages of ischemia on the ECG were not investigated so far. In contrast to e.g. [140], metabolic effects of acute cardiac ischemia were modeled only indirectly in this work in order to reduce computing time. The same reason was responsible for the use of the monodomain model, which in general is sufficient for the computation of cardiac excitation propagation and body surface ECGs [207]. However, the bidomain model could be required for the simulation of phase 1b ischemia, since the anisotropy ratio might change in this case. The influence of the width of the BZ and the location of the ischemic region were analyzed in a previous work of this author [208]. A wider BZ facilitated the conduction towards the CIZ. Different locations of the ischemic region influenced, which ECG lead presented the most pronounced ischemia effects.

Although shifts of the ST segment of the 12-lead ECG are used as a diagnostic marker of acute cardiac ischemia and myocardial infarction, a significant part of the patients showing "electrically silent" ischemia without these changes are missed [1]. As suggested e.g. in [209], the BSPM could help to identify these ischemia cases. In the present simulation study, even the BSPM of setup 3 did not show significant differences during the first 10 min compared to the control case, although this nearly complete compensation of injury currents might be due to the symmetrical shape of the ellipsoidal ischemic region. Since these cases are assumed to be unlikely, a diffusion model of the coronary blood flow as presented in e.g. [210] should be used to create more realistic ischemic regions in future simulations. Nevertheless, the use of BSPMs could improve the diagnosis of ischemia, since posterior myocardial infarctions or ischemia are hard to diagnose using the 12-lead ECG [209]. Furthermore, other markers as troponin T or CK-MB also help to reliably detect ischemia, but with a delay of typically three hours.

11

Results: Modeling the Impact of Pharmacological Therapy on Cardiac Electrophysiology

The results of multiscale simulations of the effects of pharmacological agents on cardiac electrophysiology outlined in chapter 7 and in [18] are presented in this chapter. Initially, the impact of two antiarrhythmic agents, dronedarone and amiodarone, on control and electrically remodeled human atrial electrophysiology was investigated. Furthermore, the effects of the gastroprokinetic agent cisapride and the antiarrhythmic amiodarone, which both cause QT prolongation, were analyzed in healthy and ischemic ventricular tissue.

11.1 Pharmacological Treatment of Control and Remodeled Atria

In this section, the influence of dronedarone and amiodarone on control and electrically remodeled atrial electrophysiology was investigated using the Courtemanche model. The multiscale simulations ranged from the single cell up to 2D tissue patches, in which the influence on reentrant circuits was analyzed. General differences between control and electrically remodeled atrial electrophysiology are described in section 9.2.

11.1.1 Atrial Single-Cell APs under the Influence of Dronedarone and Amiodarone

The effects of dronedarone and amiodarone on control and electrically remodeled human atrial APs are presented in Fig. 11.1. The corresponding values of the AP properties are summarized in Table 11.1. With increasing concentration of the antiarrhythmics, the AP amplitude was decreased

Table 11.1. AP properties of control and electrically remodeled atrial myocytes under the influence of dronedarone and amiodarone (BCL = 1 s). *drone:* dronedarone. *amio1:* amiodarone setup 1. *amio2:* amiodarone setup 2. *l:* low concentration. *h:* high concentration.

setup	amplitude (mV)		$V_{m,rest}$ (mV)		APD$_{50}$ (ms)		APD$_{90}$ (ms)		dV_m/dt_{max} (V/s)	
	control	*cAF*	*control*	*cAF*	*control*	*cAF*	*control*	*cAF*	*control*	*cAF*
no drug	105.6	108.0	−81.0	−84.4	172.4	87.3	298.4	147.9	190.9	194.1
drone l	103.5	106.2	−80.6	−84.3	227.3	117.3	378.2	183.5	185.3	184.3
drone h	99.6	102.6	−80.5	−84.3	248.2	127.7	398.7	191.7	163.0	154.2
amio1 l	92.2	95.1	−81.1	−84.4	230.0	109.3	359.6	165.3	131.8	122.6
amio1 h	80.6	82.7	−81.5	−84.1	218.3	142.6	334.5	187.5	66.2	65.3
amio2 l	92.6	94.9	−81.7	−84.0	114.8	93.8	261.3	147.8	124.7	116.9
amio2 h	80.6	82.7	−81.7	−84.2	122.6	98.2	241.5	143.6	67.1	66.0

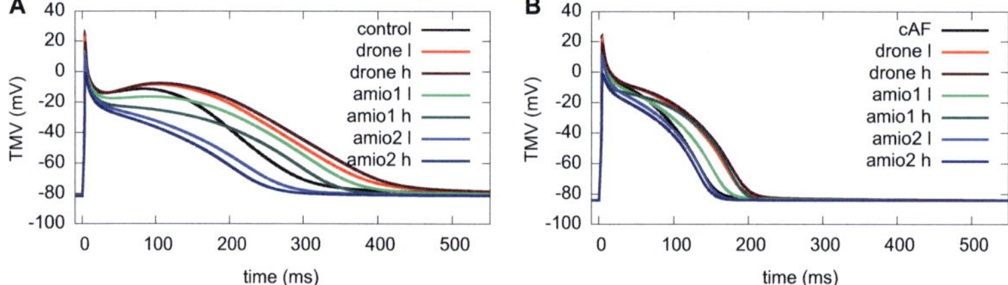

Figure 11.1. Control (A) and cAF (B) APs under the influence of dronedarone and amiodarone. *drone:* dronedarone. *amio1:* amiodarone setup 1. *amio2:* amiodarone setup 2. *l:* low concentration. *h:* high concentration.

in control and cAF myocytes in a similar fashion. However, the decrease was more pronounced in case of amiodarone (−13 to −25 mV) compared to −2 to −6 mV in case of dronedarone. Almost no difference in amplitude could be observed between setups 1 and 2. Linked to the reduced amplitude, the maximum upstroke velocity was also slowed due the administration of dronedarone and amiodarone, again to a greater extent in case of amiodarone, but also in case of cAF. The reduction of dV_m/dt_{max} was largest (−66.3%) in case of cAF-induced remodeling using the high concentration of amiodarone setup 1. In the control case, dV_m/dt_{max} was slowed by −65.3% with the same setup and concentration. Setup 2 presented an only slightly smaller decrease of dV_m/dt_{max}. The resting membrane voltage $V_{m,rest}$ showed only slight shifts in the range of −0.7 to +0.5 mV due to the pharmacological agents. Dronedarone and amiodarone setup 1 caused a significant prolongation of the APD. The high concentration of dronedarone prolonged the APD_{50} (control: $+43.9\%$, cAF: $+46.3\%$) and APD_{90} (control: $+33.6\%$, cAF: $+29.6\%$) most. In case of amiodarone setup 2, a different effect could be observed: the APD_{50} was shortened in the control case, which was more pronounced using the low concentration, whereas in case of cAF, the APD_{50} was increased with increasing concentration. However, the APD_{90} decreased with increasing concentration of amiodarone setup 2, but in case of cAF, this shortening was negligibly small (maximally -2.9%).

11.1.2 Effects of Dronedarone and Amiodarone on Restitution Properties

The impact of dronedarone and amiodarone on the restitution properties of the control and electrically remodeled 1D tissue strand is shown in Fig. 11.2. The slope of the APD_{90} curve $dAPD_{90}/dDI$, the ERP, the CV and the WL were varied over DIs between 0 and 1 s. Generally, almost no alternans could be observed in the different models, therefore, only one trace is shown for each model.

Fig. 11.2A and B present the $dAPD_{90}/dDI$, which was around zero at long DIs in all cases. The slope of the control model slightly increased to 0.3 towards a DI of 0.09 s and then decreased again. However, dronedarone caused negative values of $dAPD_{90}/dDI$ at DIs shorter than 0.09 s. In contrast, the low concentrations of amiodarone setups 1 and 2 exhibited an increase of the slope up to 1.2 at short DIs. The high concentrations of amiodarone presented similar slopes as the control case at short DIs, but also large peaks with slopes of 8.8 at a DI of 0.18 s in case of setup 1 and 2.9 at a DI of 0.33 s in case of setup 2. In case of cAF, oscillations of the slopes of the APD_{90} restitution curves to slightly negative values could be observed at short DIs. However, the $dAPD_{90}/dDI$ of the cAF model without the influence of pharmacological agents increased towards values between 0.2 and 0.4 at short DIs. Dronedarone and amiodarone setup 1 presented oscillations in a similar range, although they started already at a DI of 0.4 s. In contrast, amiodarone setup 2 caused smaller oscillations close to zero and maximally up to 0.2.

The ERP under the influence of dronedarone and amiodarone is presented in Fig. 11.2C and D. The control model caused an ERP of 0.32 s at DIs longer than 0.4 s. Then, the ERP slightly decreased to approximately 0.27 s at a DI close to 0 s. Dronedarone caused an increased ERP of around 0.4 s, which

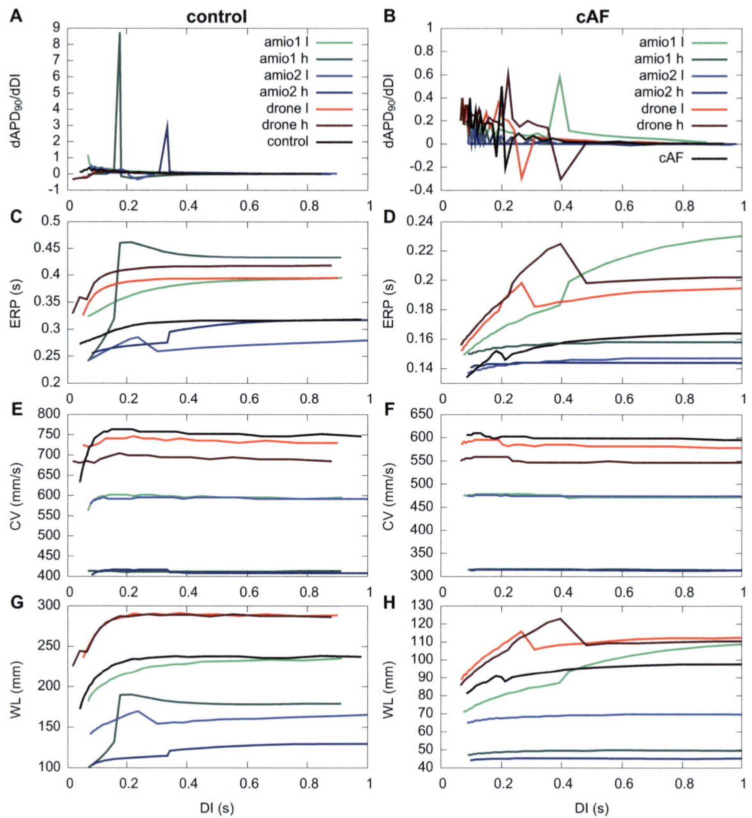

Figure 11.2. Restitution properties of control (*left*) and cAF (*right*) tissue under the influence of dronedarone and amiodarone simulated in a 1D fiber. Slope of the APD_{90} restitution (A, B), ERP (C, D), CV (E, F) and WL (G, H) were varied over the DI. *drone:* dronedarone. *amio1:* amiodarone setup 1. *amio2:* amiodarone setup 2. *l:* low concentration. *h:* high concentration.

rapidly decreased at DIs shorter than 0.2 s. With increasing concentration of dronedarone, the ERP was slightly increased. Amiodarone setup 1 presented similar ERPs at long DIs. Only the high concentration resulted in a longer ERP. The low concentration of setup 1 exhibited a slightly steeper slope than the control case, but the decrease started at DIs shorter than 0.6 s already. However, the high concentration of setup 1 caused an increase of the ERP to 0.46 s at DIs between 0.6 and 0.2 s followed by a rapid decline to 0.24 s at DIs shorter than 0.2 s. In case of setup 2, the ERP caused by the low concentration of amiodarone was around 0.28 s at long DIs and continuously decreased to a DI of 0.25 s. Towards this DI, the ERP was first increased and then decreased to 0.24 s again towards short DIs. The high concentration of amiodarone setup 2 caused a similar ERP as the control case at long DIs, but the ERP started decreasing at a DI of already 0.6 s. Furthermore, a slight discontinuous shift of the ERP towards smaller values could be observed at a DI of 0.33 s. In case of cAF, the ERP was 0.164 s at a DI of 1 s and decreased by 30 ms to short DIs. Around DIs of 0.2 s, a short increase of the ERP could be seen. Dronedarone caused an increased ERP of around 0.2 s at a DI of 1 s, which decreased to below 0.16 s at short DIs. A shift of the ERP towards higher values, which was more pronounced than in the control case, could be observed at a DI of 0.3 s in case of the low concentration and at a DI of around 0.4 s in case of the high concentration of dronedarone. The low concentration of amiodarone setup 1 exhibited the highest ERP (0.23 s) at a DI of 1 s. Towards shorter DIs , the ERP of this setup continuously decreased to 0.15 s except for a shift to smaller values at a DI of 0.4 s. Setup 2 and the high concentration of setup 1 showed smaller ERPs than the control case at long DIs, but since the values did not decrease significantly towards short DIs, slightly longer ERP values could be seen there. No significant differences between the low and high concentrations of the ERPs of setup 2 were visible.

In Fig. 11.2E and F, the CVs of the models of dronedarone and amiodarone in control and electrically remodeled 1D tissue strands are depicted. The CV of the control model was nearly 750 mm/s at a DI of 1 s. Towards a DI of 0.2 s, the CV slightly increased and afterwards rapidly decreased to 630 mm/s at short DIs. In general, the slopes of the CV restitution curves of the antiarrhythmics were flattened, so that the rate adaptation at short DIs was negligible. Increased concentration of dronedarone and amiodarone caused decreasing CVs. The high concentration of dronedarone exhibited CVs slightly below 700 mm/s. The CVs of amiodarone setups 1 and 2 were nearly the same. The low concentration of amiodarone resulted in CVs below 600 mm/s. In case of the high concentration of amiodarone, the CV was almost 200 mm/s slower. In case of cAF under the influence of antiarrhythmics, qualitatively similar results as in the control case could be obtained. The CVs of all models were shifted by around 100 to 150 mm/s towards slower values. However, the cAF-induced remodeling without the influence of antiarrhythmics showed no rate adaptation at short DIs similar to the models of antiarrhythmics.

The WL restitution of the control and electrically remodeled atrial models under the influence of dronedarone and amiodarone is given in Fig. 11.2G and H. Since the WL is the product of the ERP and the CV, the resulting curves look qualitatively similar. The control atrial model presented a WL of nearly 240 mm between DIs of 0.3 and 1 s. Below 0.3 s, the WL decreased to around 170 mm at short DIs. Independent of the concentration, the WLs were prolonged by around 50 mm in case of dronedarone. Similar WLs as in the control case could be obtained using the low concentration of amiodarone setup 1. The WL of the high concentration of amiodarone setup 1 was around 180 mm at long DIs, slightly increasing towards a DI of 0.2 s and afterwards rapidly declining to 100 mm at short DIs. Amiodarone setup 2 showed the shortest WLs, which decreased with increasing concentration. The WL of the low concentration of setup 2 decreased from more than 160 mm at a DI of 1 s to 140 ms at short DIs. The WL was reduced from 130 mm at a DI of 1 s to 100 mm at short DIs in case of the high concentration of setup 2. As in case of the ERP, discontinuous shifts of the WLs to higher and lower values could be observed in case of amiodarone setup 2 at DIs of 0.25 and 0.33 s, respectively. In case of cAF, the WL was slightly shorter than 100 mm at a DI of 1 s decreasing to around 80 mm at short DIs. The WLs of dronedarone were around 15 mm longer than in the cAF model and similar for both concentrations, except for the discontinuous shifts at 0.3 and 0.4 s, respectively. The low concentration of amiodarone setup 1 presented WLs in a similar range as the cAF model, but with a more pronounced decrease. Setup 2 and the high concentration of amiodarone setup 1 caused nearly flat WL restitution curves with values down to less than 50 mm.

Since the vulnerable window (VW) was nearly constant at all investigated DIs, only the values at a BCL of 1 s are listed in Table 11.2. Only at the shortest DI, which could be analyzed, the VW was slightly higher. In the control model without pharmacological agent, the VW was increased by increasing concentrations of dronedarone to maximally 115%. In case of low concentrations, amiodarone setup 1 resulted in a slight increase of the VW, whereas setup 2 caused a decrease of the WL by −13.8%. In case of high concentrations of amiodarone, a decrease by a maximum of −75% could be

Table 11.2. Vulnerable window (VW) of control and electrically remodeled atrial tissue under the influence of dronedarone and amiodarone (BCL $= 1$ s).

setup	VW (ms)	
	control	*cAF*
no drug	2.61	1.56
dronedarone low	2.83	1.63
dronedarone high	3.00	1.72
amiodarone setup 1 low	2.64	1.71
amiodarone setup 1 high	0.96	3.72
amiodarone setup 2 low	2.25	1.69
amiodarone setup 2 high	0.66	3.71

observed. In case of cAF without the influence of antiarrhythmics, the VW was reduced to 60% of the value of the control atrial model. All drugs increased the VW in case of cAF, especially the high concentrations of amiodarone ($+138\%$), whereas dronedarone presented moderate prolongations by up to $+10\%$.

11.1.3 Impact of Dronedarone and Amiodarone on Termination of Rotors

As described in section 7.4.2, four reentrant circuits were initiated in a 2D tissue patch using the model of cAF-induced remodeling. Then, the trajectories of the centers of the reentrant circuits were tracked during 5 s, which is shown in the upper left corner of Fig. 11.3. The spiral wave tips presented a stable curved, star-shaped meandering during the 5 s under investigation. Furthermore, a DF of 8.2 Hz was measured. The rotor initiation was repeated using the model of cAF-induced remodeling without the influence of antiarrhythmics and then, the effects of dronedarone and amiodarone were included and observed during a period of 5 s. In case of the low concentration of dronedarone, the meandering of the reentrant circuits was increased, so that the upper two rotors moved to the borders of the patch and left it after 2.25 and 2.4 s, respectively. In this case, the DF was reduced by 0.9 Hz. The high concentration also presented an increased meandering of the reentrant circuits. Due to this, the rotor in the upper left corner moved towards the border of the patch and left it after 1.9 s. However, the rotor, which was previously in the lower left corner, replaced it and moved to the upper left corner. In the meantime, a wave break occurred close to the center of the

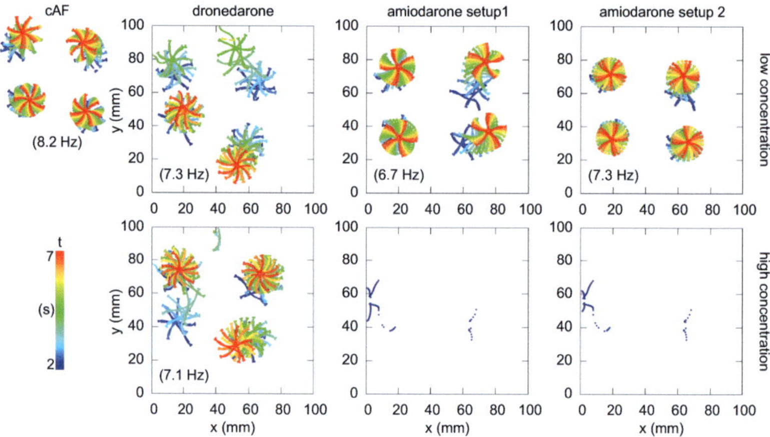

Figure 11.3. Effects of low (*top row*) and high (*bottom row*) concentrations of dronedarone and amiodarone on rotor trajectories of 4 rotors initiated in 2D cAF tissue patch. Dominant frequencies are shown in parentheses.

upper border of the patch leading to a small reentrant circuit, which was quickly eliminated by the other rotors. Finally, three reentrant circuits remained after 5 s. However, the DF was decreased by 1.1 Hz compared to the cAF case without the influence of antiarrhythmics. The low concentrations of amiodarone also increased the meandering of the rotor centers. This effect was more pronounced in setup 1, which also decreased the DF by 1.5 Hz compared to only 0.9 Hz in case of setup 2. In case of high concentrations of amiodarone, all reentrant circuits terminated after less than 100 ms. Consequently, no DF was measured in these cases.

11.2 Effects of Pharmacological Treatment on Healthy and Ischemic Ventricles

The influence of the gastroprokinetic agent cisapride and the antiarrhythmic agent amiodarone, which are both known to prolong the QT interval of the ECG [109, 211], on ventricular electrophysiology was investigated. The effects on control and stage 1 (5 min) ischemic ventricular electrophysiology were compared in a single cell, as well as in a 1D tissue patch. General differences between control and ischemic ventricular APs are described in chap-

Table 11.3. AP properties of control and ischemic (5 min after onset of effects) epicardial myocytes under the influence of cisapride and amiodarone (BCL = 1 s). *cisa:* cisapride. *amio1:* amiodarone setup 1. *amio2:* amiodarone setup 2. *l:* low concentration. *h:* high concentration.

setup	amplitude (mV)		$V_{m,rest}$ (mV)		APD$_{50}$ (ms)		APD$_{90}$ (ms)		dV_m/dt_{max} (V/s)	
	control	*5 min*	*control*	*5 min*	*control*	*5 min*	*control*	*5 min*	*control*	*5 min*
no drug	145.9	108.4	−85.6	−73.5	267.7	147.2	307.3	169.8	101.9	61.3
cisa l	146.1	108.4	−85.6	−73.5	274.7	154.4	318.4	178.2	102.0	61.2
cisa h	146.1	108.4	−85.6	−73.5	277.8	157.7	323.6	182.1	102.0	61.2
amio1 l	136.7	99.3	−85.7	−73.6	299.7	149.4	344.0	171.9	96.0	56.8
amio1 h	124.8	93.0	−85.9	−73.7	317.8	133.5	362.1	155.8	86.0	50.5
amio2 l	137.0	100.1	−86.0	−73.8	270.3	109.2	326.2	133.6	96.5	57.0
amio2 h	124.6	91.3	−86.1	−73.9	288.4	66.3	337.5	89.1	85.9	51.0

ter 10. Furthermore, the impact of the pharmacological agents on ventricular excitation propagation and the resulting body surface ECGs was analyzed using the control ventricular model.

11.2.1 Impact of Cisapride and Amiodarone on Ventricular Single-Cell APs

The effects of cisapride and amiodarone on control and ischemic (5 min) epicardial APs are presented in Fig. 11.4. In epicardial myocytes, the effects of ischemia were most pronounced (compare section 10.1) due to the increased sensitivity of $I_{K,ATP}$ to loss of ATP compared to endocardial or M cells. The effects of cisapride and amiodarone on endocardial and M cells were qualitatively similar. The resulting changes of the AP parameters of the epicardial myocytes are listed in Table 11.3. Cisapride did not alter the amplitude, resting membrane voltage and upstroke velocity of the APs compared to

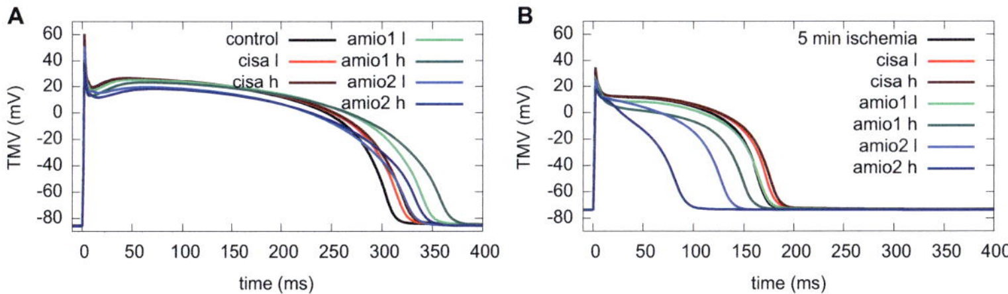

Figure 11.4. Control (A) and ischemic (B) epicardial APs under the influence of cisapride and amiodarone. *cisa:* cisapride. *amio1:* amiodarone setup 1. *amio2:* amiodarone setup 2. *l:* low concentration. *h:* high concentration. Modified according to [18].

the models without the influence of pharmacological agents, whereas amiodarone reduced the amplitude and dV_m/dt_{max} with increasing concentrations and no significant differences between setups 1 and 2. The amplitude was decreased by up to 14.6% in the control case and by 15.7% in the ischemia case using the high concentration of amiodarone. A maximal reduction of dV_m/dt_{max} by -15.7% and -17.6%, respectively, could be observed in the control and ischemia models under the influence of the high concentration of amiodarone. Amiodarone slightly shifted $V_{m,rest}$ towards more negative values by maximally $-0.5\,\text{mV}$. The APD was increased by cisapride with increasing concentration. The prolongation of the APD_{50} of the control and ischemia models due to cisapride was 7 or 10 ms depending on the concentration. In the control case, amiodarone increased the APD_{50} with increasing concentrations. Using the high concentration of setup 1, the effect was most pronounced resulting in a prolongation by around 50 ms. However, the same pharmacological agent caused a shortening of the APD_{50} in case of ischemia, whereas the low concentration presented a slight prolongation. The APD_{50} was shortened more using amiodarone setup 2. The high concentration of setup 2 caused a maximal shortening by approximately 80 ms in case of ischemia. The APD_{90} of the pharmacological agents presented qualitatively similar results: Cisapride prolonged the APD with increasing concentration in the control and ischemia cases, whereas amiodarone presented this effect only in the control cases. After 5 min of ischemia, the low concentration of setup 1 slightly prolonged the APD_{90}, whereas the high concentration shortened it. A more pronounced shortening could be seen in case of amiodarone setup 2. After 5 min of ischemia, the APD_{90} was decreased by a maximum of 70 ms using the high concentration of setup 2.

11.2.2 Restitution Curves of Ventricular Tissue under the Influence of Cisapride and Amiodarone

The restitution properties of the control and ischemia model 5 min after the onset of effects shown in Fig. 11.5 were investigated under the influence of cisapride and amiodarone in a 1D epicardial tissue strand. As in case of the single-cell analyses, the ischemia effects were strongest in this tissue. Furthermore, the impact of cisapride and amiodarone on endocardial and midmyocardial tissue was qualitatively similar. Generally, the slope of the

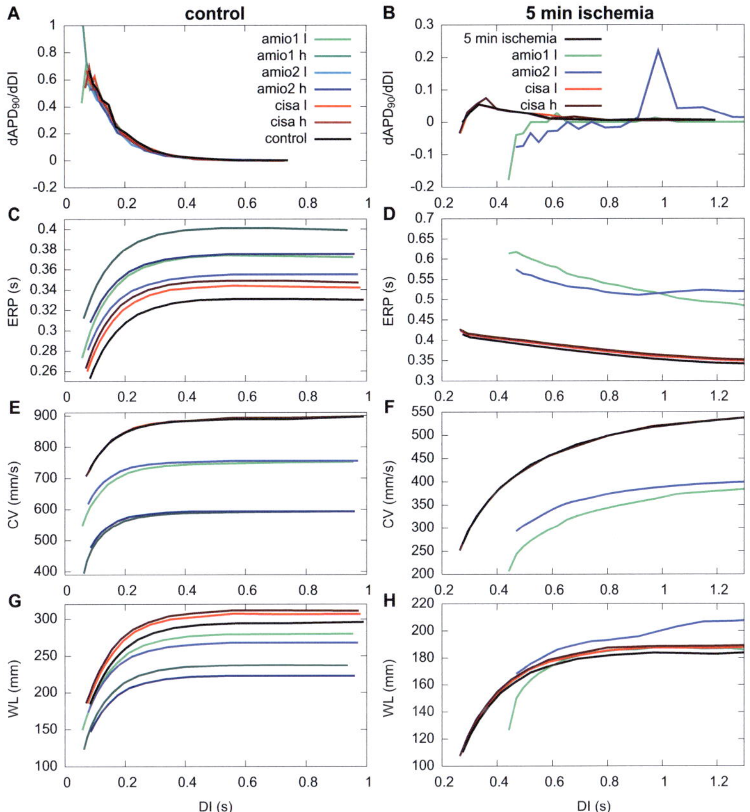

Figure 11.5. Restitution properties of control and ischemic epicardial tissue under the influence of cisapride and amiodarone simulated in a 1D fiber. Slope of the APD_{90} restitution (A, B), ERP (C, D), CV (E, F) and WL (G, H) were varied over the DI. *cisa:* cisapride. *amio1:* amiodarone setup 1. *amio2:* amiodarone setup 2. *l:* low concentration. *h:* high concentration. Modified according to [18].

APD_{90} restitution curve, the CV and the WL were reduced after 5 min of ischemia, whereas the ERP was increased, which is called postrepolarization refractoriness. The ERP increased almost linearly with decreasing DI in case of ischemia. Besides, the restitution properties of the ischemic models initiated APs at every beat at DIs down to 0.27 s in case of cisapride and to 0.45 s in case of amiodarone, whereas the control models could be investigated at DIs shorter than 60 ms. High concentrations of amiodarone did not allow excitation propagation in ischemic tissue. Additionally, almost no alternans could be observed in the different models. Therefore, only one trace is shown for each model.

The slope of the APD_{90} restitution of the models under the influence of cisapride and amiodarone is presented in Fig. 11.5A and B. At long DIs, $dAPD_{90}/dDI$ was nearly zero in all cases. In the control case, $dAPD_{90}/dDI$ increased towards shorter DIs. Without the influence of pharmacological agents, a maximum of almost 0.7 was reached. Cisapride presented similar values as the control model, whereas the high concentration of amiodarone setup 1 caused a slope of 1 at short DIs. In contrast, setup 2 and the low concentration of amiodarone setup 1 slightly decreased the slope at short DIs. In case of ischemia, the slope increased only slightly with decreasing DIs to around 0.05 and then decreased to zero again at short DIs. Cisapride presented almost the same slopes, whereas amiodarone caused small negative values with oscillations at short DIs.

The influence of cisapride and amiodarone on the ERP is shown in Fig. 11.5C and D. In the control case, the pharmacological agents did not change the slope of the ERP curves, which rapidly decreased at DIs smaller than 0.4 s. The control model exhibited an ERP of 0.33 s at long DIs, which decreased to less than 0.26 s at short DIs. Cisapride slightly increased the ERP with increasing concentration. This effect was more pronounced in case of amiodarone setup 2 and strongest in case of setup 1, which led to a maximal prolongation of 70 ms in the control case and of around 150 ms in case of ischemia. After 5 min of ischemia without the influence of cisapride or amiodarone, the ERP started at 0.35 s at a DI of 1 s and increased to more than 0.4 s at short DIs. Amiodarone setup 1 presented a steeper increase of the ERP towards short DIs than the ischemia model without the influence of pharmacological agents.

In Fig. 11.5E and F, the CV restitution of the control and ischemia models is depicted. In the control case, the CV started at around 900 mm/s at long DIs. At DIs shorter than 0.3 s, the CV declined to nearly 700 mm/s. Cisapride presented the same CV restitution. In contrast, amiodarone decreased the CV with increasing concentration. In case of the high concentration, the CV was reduced by around 300 mm/s. Compared to setup 2, setup 1 could be investigated at lower DIs leading to smaller CVs, which was the only differ-

Table 11.4. VW of control and ischemic epicardial tissue under the influence of cisapride and amiodarone (BCL = 1 s). The high concentration of amiodarone did not initiate excitation propagation in case of ischemia (marked by ×).

setup	VW (ms)	
	control	*5 min*
no drug	0.34	12.87
cisapride low	0.34	12.77
cisapride high	0.33	12.74
amiodarone setup 1 low	0.41	16.69
amiodarone setup 1 high	0.62	×
amiodarone setup 2 low	0.31	32.88
amiodarone setup 2 high	0.47	×

ence between both amiodarone setups. After 5 min of ischemia, the CV was around 500 mm/s at a DI of 1 s, which continuously decreased to 250 mm/s at short DIs. Cisapride presented exactly the same CV restitution, whereas amiodarone reduced the CV by maximally 300 mm/s. Furthermore, amiodarone setup 2 slightly flattened the slope of the CV restitution, so that the CV reduction was less pronounced at short DIs.

The impact of cisapride and amiodarone on the WL restitution is presented in Fig. 11.5G and H. The control model presented a WL of almost 300 mm at long DIs, which continuously decreased to 180 mm at DIs shorter than 0.4 s. Cisapride caused a slight prolongation of the WL with increasing concentration, whereas amiodarone reduced it with increasing concentration. The high concentration of setup 1 caused a decrease by almost 60 mm. However, a reduction by more than 70 mm could be observed in case of the high concentration of setup 2. After 5 min of ischemia, the WL was 180 mm at long DIs and reduced to 110 mm at short DIs. Cisapride caused a slight prolongation of the WL. Amiodarone setup 1 showed an increased WL at long DIs, but the WL decreased more rapidly to lower values than the model without the influence of cisapride or amiodarone resulting in decreased WLs at short DIs. Amiodarone setup 2 led to a prolongation of the WL, especially at long DIs.

As in case of the VW of dronedarone and amiodarone in atrial tissue described in section 11.1.2, the VW investigated in control and ischemic epicardial tissue under the influence of cisapride and amiodarone was almost constant at different DIs. Consequently, the VWs at a BCL of 1 s are listed in Table 11.4. Acute cardiac ischemia significantly prolonged the VW. Without the influence of pharmacological agents, the VW was almost 38 times longer after 5 min of ischemia. Cisapride on the one hand did not significantly alter the VW of the control and ischemic tissue. On the other hand, amiodarone prolonged the VW with increasing concentration except for the low concentration of setup 2, which led to a slight decrease of the VW in control tissue. In the control case, the high concentration of setup 1 showed the maximal increase by $+82\%$. During ischemia, high concentrations of amiodarone could not excite the tissue. However, the prolongation of the VW was highest ($+155\%$) using setup 2 in case of ischemia.

11.2.3 Body Surface ECGs Affected by Cisapride and Amiodarone

In order to investigate the impact of cisapride and amiodarone on the body surface ECG, the excitation propagation was computed in the control ventricular model presented in section 3.3, first. Compared to the control model without the influence of pharmacological agents, cisapride affected the ventricular repolarization only, whereas the depolarization and plateau phase remained the same. Since the APD was prolonged by cisapride, the complete repolarization of the ventricles was delayed. In contrast, amiodarone also had an impact on the depolarization of the ventricles, since the upstroke velocity and the amplitude were decreased in this case. Furthermore, the excitation propagation was slowed due to the decreased CV under the influence of amiodarone. Due to this, the complete depolarization of the ventricles was delayed and the transmembrane voltage did not reach the same amplitude as in case of the control model. Additionally, the APD was prolonged by amiodarone, so that the repolarization of the ventricles started later.

These underlying effects also caused changes in the corresponding body surface ECGs under the influence of cisapride and amiodarone. Since the signal amplitudes were highest in lead V_4, the resulting changes could be best observed there. The ECGs of the control ventricular model with and without the influence of pharmacological agents is shown in Fig. 11.6. The control model produced a narrow QRS complex and after the flat ST segment a narrow T wave ending after 405 ms, which equated to the QT interval. In case of cisapride, the QRS complex was the same as in the control model. Only the onset and end of the T wave were delayed. Consequently, the low

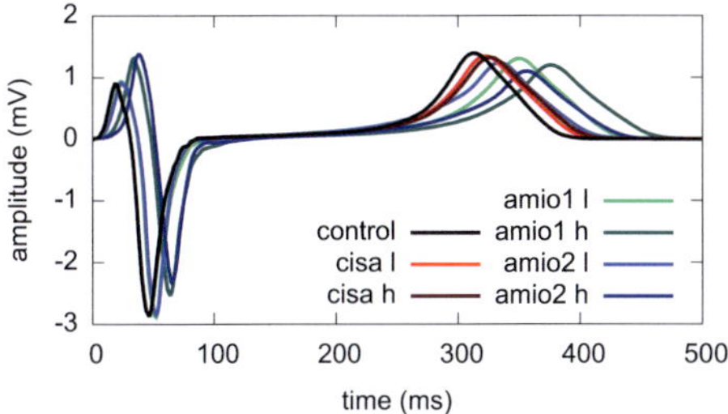

Figure 11.6. Body surface ECGs of control ventricles under the influence of cisapride and amiodarone. *cisa:* cisapride. *amio1:* amiodarone setup 1. *amio2:* amiodarone setup 2. *l:* low concentration. *h:* high concentration.

and high concentrations of cisapride prolonged the QT interval by 17 and 26 ms, respectively. Low concentrations of amiodarone presented a similar QRS complex, which was prolonged, whereas high concentrations also delayed it. Additionally, the positive peak of the QRS complex resulting from high concentrations had an increased amplitude and the negative peak a reduced amplitude instead. The onset of the T wave was delayed by amiodarone setup 1 with increasing concentration. Additionally, the T wave was broadened in this case. Due to this, the QT interval was prolonged by 47 and 75 ms, respectively. In case of the low concentration of amiodarone setup 2, broadening of the T wave could be observed leading to an increase of the QT interval by 39 ms. In contrast, the onset of the T wave was slightly delayed and the amplitude of the T wave was smaller due to the high concentration of setup 2. Therefore, the resulting QT interval was also 47 ms longer than the interval of the control model.

11.3 Discussion

In this study, the effects of three pharmacological agents, dronedarone, amiodarone and cisapride, were investigated in multiscale simulations of control atrial or ventricular tissue, as well as in models of cAF or acute cardiac ischemia. Dronedarone caused a moderate reduction of the atrial AP amplitude and upstroke velocity, as well as a prolongation of the APD. Consequently, the ERP and WL were increased. The CV was decreased in a 1D atrial tissue strand. Only in the control model, a slight reduction of dAPD$_{90}$/dDI could be observed at short DIs, whereas the slope remained similar in case of cAF. In addition, dronedarone prolonged the VW in atrial tissue. Nevertheless, the number of rotors and the DF could be reduced after the administration of dronedarone in a 2D model of cAF.

Since the reported experimental data on the pharmacological inhibition of I_{CaL} and I_{Kr} due to amiodarone covered large ranges in the literature, two setups with different reduction of the maximal channel conductivities were created. As a consequence, also some of the resulting effects of both setups varied. The effects of amiodarone were investigated in atrial and ventricular tissue, since it is used for the treatment of both, cAF and acute cardiac ischemia [85, 87]. In general, amiodarone decreased the AP amplitude and the upstroke velocity. The APD_{90} was increased by setup 1 in case of atrial and ventricular cells except for the high concentration in case of acute cardiac ischemia. A shortening of the APD_{90} could also be observed using setup 2 except for the control ventricular model, which exhibited a prolonged duration. Mainly setup 2 presented a slight flattening of the APD_{90} restitution slope, whereas it decreased the ERP in atrial tissue. Setup 1 and both ventricular setups prolonged the ERP. The CV was generally reduced in atrial and ventricular tissue with increasing concentration and almost no differences between both setups. As a consequence, amiodarone reduced the WL, except for ischemic ventricular tissue. The effects of amiodarone on the VW varied also. In most cases, the VW was prolonged, only the low concentration of setup 2 caused a slight decrease in control ventricular tissue and the high concentrations of both setups presented a significant shortening in the control atrial tissue. Nevertheless, the DF was reduced in 2D atrial models of cAF and all rotors were terminated completely using the high concentration of amiodarone. The main effect responsible for the termination of the rotors was probably the blockade of I_{Na}, which increased the core size of the reentrant circuits as described in [212]. Therefore, the existing rotors could not excite the surrounding tissue again. Amiodarone also influenced the cardiac excitation propagation in the 3D ventricular model leading to pronounced QT prolongation in the body surface ECGs.

Cisapride affected cardiac repolarization only; therefore, amplitude and upstroke velocity of the AP, as well as the CV and VW were nearly unchanged. However, APD and ERP were prolonged due to cisapride with increasing concentration, causing a slight prolongation of the WL, too. In the body surface ECGs, this was reflected by a prolongation of the QT interval. Since cisapride and amiodarone presented similar QT prolongations, this marker alone seems to be insufficient for the identification of drug-induced arrhythmias, as also stated in [213]. Therefore, different markers have to be investigated to effectively identify drug cardiotoxicity [214].

In comparison to experimental data, chronic treatment with dronedarone also increased APD_{90} and ERP in rabbit atrial tissue [106]. Furthermore, recent clinical studies showed that dronedarone was not as effective as amiodarone in the treatment of cAF. Better results could be obtained in the treatment of non-permanent AF instead [102]. The simulations shown in this study also showed that amiodarone effectively terminated all reentrant circuits, whereas dronedarone only reduced their number. Experimental data of APD under the influence of amiodarone was conflicting in the literature as well [87], presenting an increase of the APD, no change or a reduction depending on the study and the species investigated. The different results for setups 1 and 2 reflect the broad range of experimental data. The observed predominant APD prolongation in the atria compared to the ventricles [96] could not be shown in the present study. Human ventricular APs were analyzed in a small study consisting of 4 control human hearts and 5 hearts treated with amiodarone [215]. This study presented no significant change of the APD_{90}, but a slight increase of the AP amplitude and the upstroke velocity was reduced to only around 60%. This indicates that the inhibition of I_{Na} was overestimated by the underlying experimental data used in the present study. A different study [216] of patients with ventricular tachycardia, who presented a broadened QRS complex only at high pacing rates, supports this assumption. In [217], flattening of the APD restitution was identified as a possible antiarrhythmic mechanism of amiodarone, which could also be observed in the simulations shown in the present study. Experimental studies using cisapride also showed a prolongation of the APD and the QT intervals in guinea pig and rabbit [112], as well as in dog ventricular wedges [113]. Furthermore, QT prolongation and occurrence of torsades de pointes were observed in humans, which finally led to withdrawal from market [109].

In a previous work, transmurally varying effects of amiodarone and d-sotalol were simulated in a 3D canine ventricular wedge model [153]. A reduced transmural dispersion of the repolarization was found to decrease the vulnerability of this model. However, only inhibition of I_{NaL}, I_{CaL} and I_{Ks}

were considered in this study. The effects of dronedarone were simulated in [155], however, a reduction of ion channel conductivities of only I_{Na}, I_{Kr}, I_{Ks} and I_{CaL} was considered. Furthermore, only changes in single-cell APD restitution were determined. The effects of a different antiarrhythmic agent, dofetilide, on the ECG were simulated in [151]. Instead of an endo-cardial stimulation profile mimicking the Purkinje network, an apical stim-ulation was used leading to a non-physiological excitation propagation pat-tern and ECG. The effects of amiodarone on cAF were investigated in [9] using the model of cAF-induced remodeling published in [175]. The same cAF model was used in [10] to analyze the effects of dronedarone in human atrial tissue. In [94], the impact of amiodarone and cisapride on ventricular electrophysiology and the ECG was studied. However, the influence on acute cardiac ischemia and more recent measurement data were included in the present study. Further computational studies, as e.g. [132, 150, 156], investi-gated the impact of pharmacological treatment on cardiac electrophysiology. The influence of dronedarone, amiodarone and cisapride was analyzed in multiscale simulations of cardiac electrophysiology. These simulations re-vealed the slightly different electrophysiological properties of amiodarone and dronedarone, which is a derivative compound of amiodarone. Futher-more, differences at the lower simulation scales could explain the antiar-rhythmic properties of amiodarone and possible mechanisms responsible for the occurrence of torsades de pointes in case of cisapride.

Due to a lack of measurement data, experimental values of various species and temperatures were included in this study. Data obtained at body tem-perature and from different regions of the heart could better be used for the integration of transmurally varying effects of pharmacological agents. Fur-thermore, the biophysics of the ion channels could be adapted to effects of pharmacological treatment as described in chapter 8. In this way, frequency- and state-dependency of the effects could also be included in future simula-tions.

12

Conclusion and Outlook

In the course of this thesis, the effects of atrial and ventricular rhythm disorders, as well as pharmacological therapy, were investigated in multiscale simulations of human cardiac electrophysiology.

Ion Channel Level

A framework for the adaptation of model parameters describing the gating behavior of cardiac ion channels to measured voltage or patch clamp recordings was developed and implemented. Current recordings of wild type (WT) and mutated hERG and KCNA5 channels were preprocessed, so that the models of I_{Kr} and I_{Kur} could be adapted in this study. A trust-region-reflective (TRR) algorithm and two derivative-free population-based algorithms, a particle swarm optimization (PSO) and a genetic algorithm (GA), were evaluated alone and in combination regarding their ability to minimize the difference between simulated and measured ion currents. The combination of the PSO and the TRR algorithm achieved the smallest least squares. However, the GA combined with the TRR algorithm caused smaller median least squares with less deviations. Both algorithms successfully reconstructed most of the parameter values of synthetic clamp data with known parameter values. Only some parameters, which influenced the time constant of the gating process, depended on each other and showed larger deviations from the original values. However, the simulated current traces reproduced the measured clamp recordings of the WT and mutated channels well. Furthermore, the sensitivity of the function to be minimized to changes of the parameter values was analyzed around the minimum, which resulted from adaptation to the hERG WT data. The function to be minimized exhibited small sensitivity to changes of some of the parameters or even discontinuities in the defined parameter ranges. Due to this, the results of the single algorithms often depended on the initial parameter vectors showing the occurrence of local minima of this optimization problem. A comparable sensitivity analysis should also be carried out with the other clamp data used in this study, especially the I_{Kur} data. As a consequence of the sensitivity analysis, parameters with negligible influence on the result could be discarded in future parameter adaptations. In addition to I_{Kr} and I_{Kur}, other ion currents with mutations of the ion channel proteins or under the influence of pharmacological agents could be investigated in future work. For this purpose, also more sophisticated models of cardiac ion currents could be used.

Single-Cell Level

Five models of human atrial electrophysiology were benchmarked regarding general model properties, such as long term stability and alternans, as well as to their ability to reproduce effects of chronic atrial fibrillation (cAF)-induced remodeling. For this purpose, certain maximal ion channel conductivities were modified according to reported values in the literature.

The Courtemanche model was selected for further simulations of KCNA5 mutations and the effects of pharmacological treatment of AF using dronedarone and amiodarone.

In ventricular myocytes, the effects of the first 30 minutes of acute cardiac ischemia were integrated into the ten Tusscher model describing dynamic and heterogeneous changes of model parameters. Furthermore, the effects of cisapride and amiodarone were included in this model. The changes due to pharmacological agents were integrated by adaptation of the maximal conductivities of the affected ion channels.

The impact of cAF, acute cardiac ischemia, ion channel mutations and pharmacological treatment on action potential (AP) properties were analyzed. cAF reduced the AP duration (APD) and resting membrane voltage and increased the maximal upstroke velocity and amplitude, whereas the investigated I_{Kur} mutations mainly increased the transmembrane voltage during the plateau phase. With increasing duration of acute cardiac ischemia, the APs were shortened, the amplitudes and maximal upstroke velocities decreased and the resting membrane voltages elevated. Depending on the pharmacological agents, the resulting changes varied. Cisapride affected only the repolarization causing a prolongation of the APs. However, dronedarone and amiodarone influenced the depolarization and plateau phase as well, resulting in decreased amplitudes and maximal upstroke velocities of the APs. Since the reported values of the inhibition of ion channels due to amiodarone varied in the literature, two setups causing different results were investigated. If available, measurement data of the same species recorded at body temperature should be used in future work. Additionally, the impact of other pharmacological agents, especially on mutations of cardiac ion channels, should be studied.

1D Tissue Level

The restitution properties of the adapted electrophysiological models were investigated. For this purpose, restitution curves of the APD_{90}, effective refractory period (ERP), conduction velocity (CV) and wave length (WL) were computed in a 1D tissue strand. In addition, the slope of the APD_{90} restitution and the vulnerable window (VW), which describes the interval, during which a unidirectional block can be initiated, were determined in case of the pharmacological agents.

cAF caused a reduction of the APD_{90}, ERP, CV and WL in most models of human atrial electrophysiology. In case of the I_{Kur} mutations, the mainly observed effect was pronounced beat-to-beat alternans of the restitution properties. Depending on the number of stimuli, the values were either in a similar range as the control model, or significantly reduced.

During acute cardiac ischemia, the ERP was prolonged (postrepolarization refractoriness), whereas the CV and WL were decreased. Cisapride increased the ERP and did not change the CV and VW resulting in an increased WL. However, dronedarone and amiodarone reduced the CV. The ERP and therefore also the WL were prolonged by dronedarone, as well as the VW. Depending on the amiodarone setup, the ERP was either slightly increased or decreased resulting in a shortened WL in most cases. Amiodarone reduced the VW only in case of the control atrial model, but increased it otherwise. The WL and VW indicate the probability for the initiation of reentrant circuits and therefore the arrhythmic potency of models of cardiac electrophysiology in a simple 1D tissue strand. However, some effects, as e.g. the initiation of reentrant circuits at heterogeneous ischemic regions, can not be considered in these simulations. Therefore, the effects of altered electrophysiology also have to be analyzed at higher simulation levels.

2D Tissue Level

More detailed information on cardiac excitation propagation can be obtained from 2D tissue patch simulations, in which the behavior of reentrant circuits was investigated. The five different models of human atrial electrophysiology presented different spiral wave tip trajectories and dominant frequencies (DFs) after the initiation of a single rotor. The control Courtemanche and Grandi models failed to initiate a reentrant circuits, which is assumed to be the physiological case. Due to the decreased WL, all models of cAF-induced remodeling successfully initiated rotors with circular, elliptical or star-shaped meandering of the rotor centers. In case of the analyzed I_{Kur} mutations, a single rotor, which was not stable over time, could be initiated. The pronounced APD alternans caused chaotic excitation propagation patterns including wave breaks showing the proarrhythmic properties of these mutations. Furthermore, the impact of dronedarone and amiodarone on the termination of cAF was investigated. Four rotors were initiated in the 2D patch without the influence of pharmacological agents. After inclusion of the effects of dronedarone and amiodarone, the DFs and rotor center trajectories were analyzed. Dronedarone reduced the number of rotors and the DF, whereas amiodarone even terminated all rotors at the high concentration. Nevertheless, the effects of atrial fibrillation should be simulated in realistic 3D geometries of the human atria in future work. There, anatomical influences might also provide new findings regarding the initiation and termination of AF.

3D Tissue Level

In 3D simulations of ventricular excitation propagation, the influence of acute cardiac ischemia and pharmacological treatment with cisapride and amiodarone was studied. In the ischemic regions, ischemia effects worsened with increasing time after the onset resulting in conduction block in the CIZ at worst. Increasing transmural extent of the ischemic region from subendocardial to transmural increased the impact, too, since the effects were most pronounced in epicardial tissue. Cisapride slowed the repolarization of the ventricles, whereas amiodarone also delayed the complete depolarization of the ventricles due to the reduced CV. Finally, the changes of the ventricular excitation propagation were forward calculated on the body surface, where virtual electrocardiograms (ECGs) were derived from the extracellular potentials. Acute cardiac ischemia led to slight ST segment depression in case of subendocardial ischemia and to pronounced ST segment elevation in case of transmural ischemia in leads close to the ischemic region. In between, "electrically silent" ischemia cases presenting almost no changes in the ECG could be observed. Therefore, the diagnosis of acute cardiac ischemia should always be based on more markers, as e.g. the troponin T level.

Cisapride and amiodarone significantly prolonged the QT interval. Additionally, amiodarone broadened the QRS complex due to the reduced CV. As a consequence, prolongations of the QT interval do not necessarily indicate proarrhythmic properties of a pharmacological compound as in case of cisapride, since amiodarone is used as an antiarrhythmic agent. In this case, different markers at the lower simulation scales give information about the pro- or antiarrhythmic potency of a compound. In future work, more ischemic regions should be simulated at different occlusion sites with realistic shapes derived from diffusion models of blood flow in the coronary arteries. Furthermore, the simulations of acute cardiac ischemia should be compared to clinically measured ECGs and ischemic regions, if the data are available. In addition, transmurally varying effects of pharmacological agents should be analyzed and possibly compared to measured ECGs.

Summing up, the results presented in this thesis help to understand the underlying mechanisms responsible for the initiation or termination of rhythm disorders. The multiscale simulations revealed the complex impact of pharmacological agents, which showed anti- as well as proarrhythmic properties depending on the analyzed aspect. The patient-specific pharmacological therapy might be improved especially by the investigation of genetic defects of cardiac ion channels as presented in this work.

List of Figures

List of Publications and Supervised Theses

Journal Articles

1. **M. Wilhelms**, A. Loewe, J. Schmid, M. J. Krause, F. Fischer, E. P. Scholz, O. Dössel, and G. Seemann, "Efficient Adaptation of an Ion Current Model of Cardiac Electrophysiology to Voltage Clamp Recordings," *(in preparation)*, 2013
2. **M. Wilhelms**, H. Hettmann, M. M. C. Maleckar, J. T. Koivumäki, O. Dössel, and G. Seemann, "Benchmarking electrophysiological models of human atrial myocytes," *Front. Physiol.*, vol. 3, 2013.
3. **M. Wilhelms**, C. Rombach, E. P. Scholz, O. Dössel, and G. Seemann, "Impact of amiodarone and cisapride on simulated human ventricular electrophysiology and electrocardiograms," *Europace*, vol. 14, pp. v90–v96, 2012.
4. **M. Wilhelms**, O. Dössel, and G. Seemann, "In silico investigation of electrically silent acute cardiac ischemia in the human ventricles," *IEEE Trans. Biomed. Eng.*, vol. 58, pp. 2961–2964, 2011.
5. O. Dössel, M. W. Krueger, F. M. Weber, **M. Wilhelms**, and G. Seemann, "Computational modeling of the human atrial anatomy and electrophysiology," *Med. Biol. Eng. Comput.*, vol. 50, pp. 773–799, 2012.
6. G. Seemann, D. U. J. Keller, M. W. Krueger, F. M. Weber, **M. Wilhelms**, and O. Dössel, "Electrophysiological modeling for cardiology: methods and potential applications," *Inf. Technol.*, vol. 52, pp. 242–249, 2010.

Conference Contributions

1. **M. Wilhelms**, J. Schmid, M. J. Krause, N. Konrad, J. Maier, E. P. Scholz, V. Heuveline, O. Dössel, and G. Seemann, "Calibration of Human Cardiac Ion Current Models to Patch Clamp Measurement Data," in *Computing in Cardiology*, vol. 39, pp. 229–232, 2012.
2. **M. Wilhelms**, O. Dössel, and G. Seemann, "Comparing Simulated Electrocardiograms of Different Stages of Acute Cardiac Ischemia," in *FIMH 2011, LNCS*, vol. 6666, pp. 11–19, 2011.
3. **M. Wilhelms**, O. Dössel, and G. Seemann, "Simulating the Impact of the Transmural Extent of Acute Ischemia on the Electrocardiogram," in *Computing in Cardiology*, vol. 37, pp. 13–16, 2010.
4. **M. Wilhelms**, G. Seemann, M. Weiser, and O. Dössel, "Benchmarking Solvers of the Monodomain Equation in Cardiac Electrophysiological Modeling" in *Biomed. Tech.*, vol. 55, pp. 99–102, 2010.
5. **M. Wilhelms**, G. Seemann, and O. Dössel, "Benchmarking different models describing sinus node heterogeneity" in *IFMBE Proc.*, vol. 22, pp. 2691–2694, 2008.
6. A. Loewe, W. H. W. Schulze, Y. Jiang, **M. Wilhelms**, and O. Dössel, "Determination of optimal electrode positions of a wearable ECG monitoring system for detection of myocardial ischemia: a simulation study," in *Computing in Cardiology*, vol. 38, pp. 741–744, 2011.
7. G. Seemann, P. Carrillo Bustamante, S. Ponto, **M. Wilhelms**, E. P. Scholz, and O. Dössel, "Atrial fibrillation-based electrical remodeling in a computer model of the human atrium," in *Computing in Cardiology*, vol. 37, pp. 417–420, 2010.

Other Contributions

1. **M. Wilhelms**, "Analysis of cardiac ischemia regarding ECG and stability in a computer model of the human ventricles" Diploma Thesis, Institute of Biomedical Engineering, Universität Karlsruhe (TH), 2009.
2. **M. Wilhelms**, "Benchmarking different models describing sinus node heterogeneity" Student Research Project, Institute of Biomedical Engineering, Universität Karlsruhe (TH), 2008.

3. G. Seemann, M. W. Krueger, and **M. Wilhelms**, *Der Virtuelle Patient*, vol. 16, ch. Elektrophysiologische Modellierung und Virtualisierung für die Kardiologie – Methoden und potenzielle Anwendungen. Dresden: Health Academy, 2012.

Supervised Theses

1. Stefan Ponto, "Effects of Amiodarone on Chronic Atrial Fibrillation: A Simulation Study in Human Cells and Tissue", Student Research Project, 2010
2. Axel Loewe, "Comparison of Cardiac Simulation Tools Regarding the Modeling of Acute Ischemia", Bachelor Thesis, 2010
3. Lukas Holl, "Modeling the Impact of Dronedarone on Human Atrial Electrophysiology in Healthy and Electrically Remodeled Tissue", Student Research Project, 2011
4. Franziska Grimm, "Modeling and Simulating the Electrophysiological Effects of Phase 1b Ischemia", Student Research Project, 2011
5. Hanne Hettmann, "Benchmarking Electrophysiological Models of Human Atrial Myocytes", Bachelor Thesis, 2011
6. Niko Konrad, "Methods for the Integration of Ionic Current Measurement Data into Models of Cardiac Electrophysiology", Bachelor Thesis, 2012
7. Jochen Schmid, "Optimierte Parameteranpassung für Simulationen der menschlichen Vorhofelektrophysiologie", Diploma Thesis, 2012
8. Vanessa Lupici-Baltzer, "Automatic Initiation and Investigation of Atrial Fibrillation in Electrically Remodeled Tissue", Bachelor Thesis, 2012
9. Axel Loewe, "Arrhythmic Potency of Human Electrophysiological Models Adapted to Chronic and Familial Atrial Fibrillation", Master Thesis, 2013 (*in preparation*)

References

[1] C. Owens, A. McClelland, S. Walsh, B. Smith, and J. Adgey, "Comparison of value of leads from body surface maps to 12-lead electrocardiogram for diagnosis of acute myocardial infarction," *Am. J. Cardiol.*, vol. 102, pp. 257–265, 2008.

[2] N. Konrad, "Methoden für die Integration von Ionenstrommessdaten in Modelle kardialer Elektrophysiologie," Bachelor Thesis, Institute of Biomedical Engineering, Karlsruhe Institute of Technology (KIT), Karlsruhe, 2012.

[3] J. Schmid, "Optimierte Parameteranpassung für Simulationen der menschlichen Vorhofelektrophysiologie," Diploma Thesis, Institute of Biomedical Engineering, Karlsruhe Institute of Technology (KIT), Karlsruhe, 2012.

[4] A. Loewe, "Arrhythmic Potency of Human Electrophysiological Models Adapted to Chronic and Familial Atrial Fibrillation," Master Thesis, Institute of Biomedical Engineering, Karlsruhe Institute of Technology (KIT), Karlsruhe, 2013 (*in preparation*).

[5] H. Hettmann, "Benchmarking electrophysiological models of human atrial myocytes," Bachelor Thesis, Institute of Biomedical Engineering, Karlsruhe Institute of Technology (KIT), Karlsruhe, 2012.

[6] V. Lupici-Baltzer, "Automatic initiation and investigation of atrial fibrillation in electrically remodeled tissue," Bachelor Thesis, Institute of Biomedical Engineering, Karlsruhe Institute of Technology (KIT), Karlsruhe, 2012.

[7] A. Loewe, "Comparison of cardiac simulation tools regarding the modeling of acute ischemia," Bachelor Thesis, Institute of Biomedical Engineering, Karlsruhe Institute of Technology (KIT), Karlsruhe, 2010.

[8] F. Grimm, "Modeling and simulating the electrophysiological effects of phase 1b ischemia," Studienarbeit, Institute of Biomedical Engineering, Karlsruhe Institute of Technology (KIT), Karlsruhe, 2011.

[9] S. Ponto, "Effects of amiodarone on chronic atrial fibrillation: A simulation study in human cells and tissue," Student Research Project, Institute of Biomedical Engineering, Karlsruhe Institute of Technology (KIT), Karlsruhe, 2010.

[10] L. Holl, "Modeling the Impact of Dronedarone on Human Atrial Electrophysiology in Healthy and Electrically Remodeled Tissue," Student Research Project, Institute of Biomedical Engineering, Karlsruhe Institute of Technology (KIT), Karlsruhe, 2011.

[11] M. Wilhelms, J. Schmid, M. J. Krause, N. Konrad, J. Maier, E. P. Scholz, V. Heuveline, O. Dössel, and G. Seemann, "Calibration of Human Cardiac Ion Current Models to Patch Clamp Measurement Data," in *Computing in Cardiology*, vol. 39, (Kraków, Poland), pp. 229–232, 2012.

[12] M. Wilhelms, A. Loewe, J. Schmid, M. J. Krause, F. Fischer, E. P. Scholz, O. Dössel, and G. Seemann, "Efficient Adaptation of an Ion Current Model of Cardiac Electrophysiology to Voltage Clamp Recordings," *(in preparation)*, 2013.

[13] M. Wilhelms, H. Hettmann, M. M. C. Maleckar, J. T. Koivumäki, O. Dössel, and G. Seemann, "Benchmarking electrophysiological models of human atrial myocytes," *Front. Physiol.*, vol. 3, 2013.

[14] A. Loewe, W. H. W. Schulze, Y. Jiang, M. Wilhelms, and O. Dössel, "Determination of optimal electrode positions of a wearable ECG monitoring system for detection of myocardial ischemia: a simulation study," in *Computing in Cardiology*, vol. 38, pp. 741–744, 2011.

[15] M. Wilhelms, O. Dössel, and G. Seemann, "Simulating the Impact of the Transmural Extent of Acute Ischemia on the Electrocardiogram," in *Computing in Cardiology*, vol. 37, pp. 13–16, 2010.

[16] M. Wilhelms, O. Dössel, and G. Seemann, "Comparing Simulated Electrocardiograms of Different Stages of Acute Cardiac Ischemia," in *Functional Imaging and Modeling of the Heart* (D. N. Metaxas and L. Axel, eds.), vol. 6666 of *LNCS*, pp. 11–19, Springer, 2011.

[17] M. Wilhelms, O. Dössel, and G. Seemann, "In silico investigation of electrically silent acute cardiac ischemia in the human ventricles," *IEEE Trans. Biomed. Eng.*, vol. 58, pp. 2961–2964, 2011.

[18] M. Wilhelms, C. Rombach, E. P. Scholz, O. Doessel, and G. Seemann, "Impact of amiodarone and cisapride on simulated human ventricular electrophysiology and electrocardiograms," *Europace*, vol. 14, pp. v90–v96, 2012.

[19] L. Sherwood, *Human Physiology - From Cells to Systems*, ch. 9, Cardiac Physiology. Belmont, California: Thomson Brooks/Cole, 6 ed., 2007.

[20] G. Tortora and M. Nielsen, *Principles of Human Anatomy*, ch. 14, The Cardiovascular System: The Heart. John Wiley & Sons, Inc., 11 ed., 2009.

[21] C. S. Henriquez, A. L. Muzikant, and C. K. Smoak, "Anisotropy, fiber curvature, and bath loading effects on activation in thin and thick cardiac tissue preparations: simulations in a three-dimensional bidomain model," *J. Cardiovasc.Electrophysiol.*, vol. 7, pp. 424–444, 1996.

[22] D. D. Streeter, *Handbook of Physiology. Section 2: The Cardiovascular System*, vol. I. The Heart, ch. Gross morphology and fiber geometry of the heart, pp. 61–112. Baltimore: Williams and Wilkins: American Physiology Soci15ty, 1979.

[23] D. U. J. Keller, "Detailed anatomical and electrophysiological modeling of human ventricles based on diffusion tensor MRI," Diplomarbeit, Institute of Biomedical Engineering, Universität Karlsruhe (TH), Karlsruhe, 2006.

[24] D. M. Roden, J. R. Balser, A. L. J. George, and M. E. Anderson, "Cardiac ion channels," *Ann. Rev. Physiol.*, vol. 64, pp. 431–475, 2002.

[25] L. Sherwood, *Human Physiology - From Cells to Systems*, ch. 3, The Plasma Membrane and Membrane Voltage. Belmont, California: Thomson Brooks/Cole, 6 ed., 2007.

[26] R. F. Schmidt, *Physiologie kompakt*, ch. Erregungsphysiologie des Herzens, pp. 197–204. Berlin, Heidelberg, New York: Springer, 1999.

[27] J. Malmivuo and R. Plonsey, *Bioelectromagnetism*, ch. 4, Active Behaviour of the Membrane, pp. 66–105. New York; Oxford: Oxford University Press, 1995.

[28] K. Cole, "Dynamic electrical characteristics of the squid axon membrane," *Arch. Sci. Physiol*, vol. 3, pp. 253–258, 1949.

[29] G. Marmont, "Studies on the axon membrane; a new method," *J. Cell. Physiol.*, vol. 34, pp. 351–382, 1949.

[30] A. L. Hodgkin and A. F. Huxley, "A quantitative description of membrane current and its application to conduction and excitation in nerve," *J. Physiol.*, vol. 117, pp. 500–544, 1952.

[31] E. Neher and B. Sakmann, "Single-channel currents recorded from membrane of denervated frog muscle fibres," *Nature*, vol. 260, pp. 799–802, 1976.

[32] A. Sherman, A. Shrier, and E. Cooper, "Series resistance compensation for whole-cell patch-clamp studies using a membrane state estimator," *Biophys. J.*, vol. 77, pp. 2590–2601, 1999.

[33] S. Poelzing and D. S. Rosenbaum, "Nature, significance, and mechanisms of electrical heterogeneities in ventricle," *Anat. Rec. A Discov. Mol. Cell. Evol. Biol.*, vol. 280, pp. 1010–1017, 2004.

[34] J. Malmivuo and R. Plonsey, *Bioelectromagnetism*, ch. The Heart, pp. 119–130. New York; Oxford: Oxford University Press, 1995.

[35] P. Kirchhof, J. Bax, C. Blomstrom-Lundquist, H. Calkins, A. Camm, R. Cappato, F. Cosio, H. Crijns, H.-C. Diener, A. Goette, C. Israel, K.-H. Kuck, G. Lip, S. Nattel, R. Page, U. Ravens, U. Schotten, G. Steinbeck, P. Vardas, A. Waldo, K. Wegscheider, S. Willems, and G. Breithardt, "Early and comprehensive management of atrial fibrillation: executive summary of the proceedings from the 2nd AFNET-EHRA consensus conference 'research perspectives in AF'," *Eur. Heart J.*, vol. 30, pp. 2969–77c, 2009.

[36] F. D. Hobbs, D. A. Fitzmaurice, J. Mant, E. Murray, S. Jowett, S. Bryan, J. Raftery, M. Davies, and G. Lip, "A randomised controlled trial and cost-effectiveness study of systematic screening (targeted and total population screening) versus routine practice for the detection of atrial fibrillation in people aged 65 and over. The SAFE study," *Health Technol Assess*, vol. 9, pp. 1–74, Oct 2005.

[37] J. Xiao, D. Liang, and Y.-H. Chen, "The genetics of atrial fibrillation: from the bench to the bedside," *Annu. Rev. Genomics Hum. Genet.*, vol. 12, pp. 73–96, 2011.

[38] Y. Miyasaka, M. E. Barnes, B. J. Gersh, S. S. Cha, J. B. Seward, K. R. Bailey, T. Iwasaka, and T. S. M. Tsang, "Time trends of ischemic stroke incidence and mortality in patients diagnosed with first atrial fibrillation in 1980 to 2000: report of a community-based study," *Stroke*, vol. 36, pp. 2362–2366, 2005.

[39] R. Nieuwlaat, L. Eurlings, J. Cleland, S. Cobbe, P. Vardas, A. Capucci, J. Lopez-Sendon, J. Meeder, Y. Pinto, and H. Crijns, "Atrial fibrillation and heart failure in cardiology practice: reciprocal impact and combined management from the perspective of atrial fibrillation: results of the Euro Heart Survey on atrial fibrillation," *J. Am. Coll. Cardiol.*, vol. 53, pp. 1690–1698, 2009.

[40] A. Waldo, "Mechanisms of atrial fibrillation," *J. Cardiovasc. Electrophysiol.*, vol. 14, pp. S267–74, 2003.

[41] A. Li and E. Behr, "Advances in the management of atrial fibrillation," *Clin. Med.*, vol. 12, pp. 544–552, 2012.

[42] U. Schotten, S. Verheule, P. Kirchhof, and A. Goette, "Pathophysiological mechanisms of atrial fibrillation: a translational appraisal," *Physiol. Rev.*, vol. 91, pp. 265–325, 2011.

[43] C. Kerr, K. Humphries, M. Talajic, G. Klein, S. Connolly, M. Green, J. Boone, R. Sheldon, P. Dorian, and D. Newman, "Progression to chronic atrial fibrillation after the initial diagnosis of paroxysmal atrial fibrillation: results from the Canadian Registry of Atrial Fibrillation," *Am. Heart J.*, vol. 149, pp. 489–496, 2005.

[44] M. C. E. F. Wijffels, C. J. H. J. Kirchhof, R. Dorland, and M. A. Allessie, "Atrial Fibrillation Begets Atrial Fibrillation," *Circ.*, vol. 92, pp. 1954–1968, 1995.

[45] R. F. Bosch, X. Zeng, J. B. Gramer, K. Popovic, C. Mewis, and V. Kühlkamp, "Ionic mechanisms of electrical remodeling in human atrial fibrillation," *Cardiovasc. Res.*, vol. 44, pp. 121–131, 1999.

[46] D. R. van Wagoner, A. L. Pond, and P. M. McCarthy, "Outward K+ current densities and Kv1.5 expression are reduced in chronic human atrial fibrillation," *Circ. Res.*, vol. 80, pp. 772–781, 1997.

[47] D. Dobrev, E. Graf, E. Wettwer, H. M. Himmel, O. Hala, C. Doerfel, T. Christ, S. Schuler, and U. Ravens, "Molecular basis of downregulation of G-protein-coupled inward rectifying K+ current (IK,ACh) in chronic human atrial fibrillation: Decrease in GIRK4 mRNA cor-

relates with reduced IK,ACh and muscarinic receptor-mediated shortening of action potentials," *Circ.*, vol. 104, pp. 2551–2557, 2001.

[48] H. S. Duffy and A. L. Wit, "Is there a role for remodeled connexins in AF?: No simple answers," *J. Mol. Cell. Cardiol.*, vol. 44, pp. 4–13, 2008.

[49] M. Wilhelm, W. Kirste, S. Kuly, K. Amann, W. Neuhuber, M. Weyand, W. G. Daniel, and C. Garlichs, "Atrial distribution of connexin 40 and 43 in patients with intermittent, persistent, and postoperative atrial fibrillation," *Heart Lung Circ.*, vol. 15, pp. 30–37, 2006.

[50] L. Chen, K. Herron, B. Tai, and T. Olson, "Lone atrial fibrillation: influence of familial disease on gender predilection," *J. Cardiovasc. Electrophysiol.*, vol. 19, pp. 802–806, 2008.

[51] T. S. Potpara and G. Y. Lip, "Lone atrial fibrillation: what is known and what is to come," *Int. J. Clin. Pract.*, vol. 65, no. 4, pp. 446–457, 2011.

[52] M. J. Janse and A. L. Wit, "Electrophysiological mechanisms of ventricular arrhythmias resulting from myocardial ischemia and infarction," *Physiol. Rev.*, vol. 69, pp. 1049–1169, 1989.

[53] E. Carmeliet, "Cardiac ionic currents and acute ischemia: from channels to arrhythmias," *Physiol. Rev.*, vol. 79, pp. 917–1017, 1999.

[54] A. A. Wilde and G. Aksnes, "Myocardial potassium loss and cell depolarisation in ischaemia and hypoxia," *Cardiovasc. Res.*, vol. 29, pp. 1–15, 1995.

[55] E. Marban, M. Kitakaze, Y. Koretsune, D. T. Yue, V. P. Chacko, and M. M. Pike, "Quantification of $[Ca^{2+}]_i$ in perfused hearts. Critical evaluation of the 5F-BAPTA and nuclear magnetic resonance method as applied to the study of ischemia and reperfusion," *Circ Res*, vol. 66, pp. 1255–1267, 1990.

[56] W. t. Smith, W. Fleet, T. Johnson, C. Engle, and W. Cascio, "The Ib phase of ventricular arrhythmias in ischemic in situ porcine heart is related to changes in cell-to-cell electrical coupling. Experimental Cardiology Group, University of North Carolina," *Circ.*, vol. 92, pp. 3051–3060, 1995.

[57] J. Weiss, N. Venkatesh, and S. Lamp, "ATP-sensitive K^+ channels and cellular K^+ loss in hypoxic and ischaemic mammalian ventricle," *J. Physiol.*, vol. 447, pp. 649–673, 1992.

[58] K. Shivkumar, N. Deutsch, S. Lamp, K. Khuu, J. Goldhaber, and J. Weiss, "Mechanism of hypoxic K^+ loss in rabbit ventricle," *J. Clin. Invest.*, vol. 100, pp. 1782–1788, 1997.

[59] A. Wilde, R. Peters, and M. Janse, "Catecholamine release and potassium accumulation in the isolated globally ischemic rabbit heart," *J. Mol. Cell. Cardiol.*, vol. 20, pp. 887–896, 1988.

[60] G. J. Grover and K. D. Garlid, "ATP-Sensitive potassium channels: a review of their cardioprotective pharmacology," *J. Mol. Cell. Cardiol.*, vol. 32, pp. 677–695, 2000.

[61] S. Dhein, "Cardiac ischemia and uncoupling: gap junctions in ischemia and infarction," *Adv. Cardiol.*, vol. 42, pp. 198–212, 2006.

[62] J. R. d. Groot and R. Coronel, "Acute ischemia-induced gap junctional uncoupling and arrhythmogenesis," *Cardiovasc. Res.*, vol. 62, pp. 323–334, 2004.

[63] M. J. Janse, J. Cinca, H. Morena, J. W. Fiolet, A. G. Kleber, G. P. de Vries, A. E. Becker, and D. Durrer, "The "border zone" in myocardial ischemia. An electrophysiological, metabolic, and histochemical correlation in the pig heart," *Circ. Res.*, vol. 44, pp. 576–588, 1979.

[64] R. Coronel, J. W. Fiolet, F. J. Wilms-Schopman, A. F. Schaapherder, T. A. Johnson, L. S. Gettes, and M. J. Janse, "Distribution of extracellular potassium and its relation to electrophysiologic changes during acute myocardial ischemia in the isolated perfused porcine heart," *Circ.*, vol. 77, pp. 1125–1138, 1988.

[65] R. Coronel, F. J. Wilms-Schopman, J. W. Fiolet, T. Opthof, and M. J. Janse, "The relation between extracellular potassium concentration and pH in the border zone during regional ischemia in isolated porcine hearts," *J. Mol. Cell. Cardiol.*, vol. 27, pp. 2069–2073, 1995.

[66] P. Colonna, C. Cadeddu, R. Montisci, L. Chen, L. Meloni, and S. Iliceto, "Transmural heterogeneity of myocardial contraction and ischemia. Diagnosis and clinical implications," *Ital. Heart J.*, vol. 1, pp. 174–183, 2000.

[67] K. A. Reimer, J. E. Lowe, M. M. Rasmussen, and R. B. Jennings, "The wavefront phenomenon of ischemic cell death. 1. Myocardial infarct size vs duration of coronary occlusion in dogs," *Circ.*, vol. 56, pp. 786–794, 1977.

[68] T. Furukawa, S. Kimura, N. Furukawa, A. Bassett, and R. Myerburg, "Role of cardiac ATP-regulated potassium channels in differential responses of endocardial and epicardial cells to ischemia," *Circ. Res.*, vol. 68, pp. 1693–1702, 1991.

[69] A. L. Goldberger, *Clinical Electrocardiography: A Simplified Approach*, ch. Myocardial Ischemia and Infarction, Section II — ST Segment Depression Ischemia and Non-Q Wave Infarct Patterns. Philadelphia: Elsevier Saunders Mosby, 8 ed., 2012.

[70] J. W. Hurst, "Thoughts about the abnormalities in the electrocardiogram of patients with acute myocardial infarction with emphasis on a more accurate method of interpreting ST-segment displacement: part I," *Clin. Cardiol.*, vol. 30, pp. 381–390, Aug 2007.

[71] J. W. Hurst, "Thoughts about the abnormalities in the electrocardiogram of patients with acute myocardial infarction with emphasis on a more accurate method of interpreting S-T segment displacement: part II," *Clin. Cardiol.*, vol. 30, pp. 443–449, Sep 2007.

[72] G. Wagner, P. Macfarlane, H. Wellens, M. Josephson, A. Gorgels, D. Mirvis, O. Pahlm, B. Surawicz, P. Kligfield, R. Childers, *et al.*, "AHA/ACCF/HRS recommendations for the standardization and interpretation of the electrocardiogram: part VI: acute ischemia/infarction: a scientific statement from the American Heart Association Electrocardiography and Arrhythmias Committee, Council on Clinical Cardiology; the American College of Cardiology Foundation; and the Heart Rhythm Society. Endorsed by the International Society for Computerized Electrocardiology," *J. Am. Coll. Cardiol.*, vol. 53, pp. 1003–1011, 2009.

[73] D. Foster, *Twelve-lead electrocardiography: theory and interpretation*, ch. Ischemia and Anginal Syndromes. Springer, 2007.

[74] E. M. Vaughan Williams, "A classification of antiarrhythmic actions reassessed after a decade of new drugs," *J. Clin. Pharmacol.*, vol. 24, pp. 129–147, 1984.

[75] S. Lazareno, "Quantification of Receptor Interactions Using Binding Methods," in *Receptor-based drug design* (P. Leff, ed.), vol. 89 of *Drugs and the Pharmaceutical Sciences*, pp. 49–78, New York: Marcel Dekker Inc., 1998.

[76] A. Hill, "The possible effects of the aggregation of the molecules of haemoglobin on its dissociation curves," *J. Physiol.*, vol. 40, no. 4, pp. 4–7, 1910.

[77] C. H. Follmer, M. Aomine, J. Z. Yeh, and D. H. Singer, "Amiodarone-induced block of sodium current in isolated cardiac cells," *J. Pharmacol. Exp. Ther.*, vol. 243, pp. 187–194, 1987.

[78] D. F. Gray, A. S. Mihailidou, P. S. Hansen, K. A. Buhagiar, N. L. Bewick, H. H. Rasmussen, and D. W. Whalley, "Amiodarone inhibits the Na(+)-K+ pump in rabbit cardiac myocytes after acute and chronic treatment," *J. Pharmacol. Exp. Ther.*, vol. 284, pp. 75–82, 1998.

[79] L. Wu, S. Rajamani, J. C. Shryock, H. Li, J. Ruskin, C. Antzelevitch, and L. Belardinelli, "Augmentation of late sodium current unmasks the proarrhythmic effects of amiodarone," *Cardiovasc. Res.*, vol. 77, pp. 481–488, 2008.

[80] J. M. Ridley, J. T. Milnes, H. J. Witchel, and J. C. Hancox, "High affinity HERG K(+) channel blockade by the antiarrhythmic agent dronedarone: resistance to mutations of the S6 residues Y652 and F656," *Biochem. Biophys. Res. Commun.*, vol. 325, pp. 883–891, 2004.

[81] D. Zankov, W. Ding, H. Matsuura, and M. Horie, "Open-state unblock characterizes acute inhibition of IKs potassium current by amiodarone in guinea pig ventricular myocytes," *J. Cardiovasc. Electrophysiol.*, vol. 16, pp. 314–322, 2005.

[82] M. Nishimura, C. H. Follmer, and D. H. Singer, "Amiodarone blocks calcium current in single guinea pig ventricular myocytes," *J. Pharmacol. Exp. Ther.*, vol. 251, pp. 650–659, 1989.

[83] A. Varro, L. Virag, and J. G. Papp, "Comparison of the chronic and acute effects of amiodarone on the calcium and potassium currents in rabbit isolated cardiac myocytes," *B. J. Pharmacol.*, vol. 117, pp. 1181–1186, 1996.

[84] Y. Watanabe and J. Kimura, "Inhibitory effect of amiodarone on Na(+)/Ca(2+) exchange current in guinea-pig cardiac myocytes," *B. J. Pharmacol.*, vol. 131, pp. 80–84, 2000.

[85] F. Van de Werf, J. Bax, A. Betriu, C. Blomstrom-Lundqvist, F. Crea, V. Falk, G. Filippatos, K. Fox, K. Huber, A. Kastrati, A. Rosengren, P. Steg, M. Tubaro, F. Verheugt, F. Weidinger, M. Weis, and ESC

Committee for Practice Guidelines (CPG), "Management of acute myocardial infarction in patients presenting with persistent ST-segment elevation: the Task Force on the Management of ST-Segment Elevation Acute Myocardial Infarction of the European Society of Cardiology," 2008.

[86] A. Camm, G. Lip, R. De Caterina, I. Savelieva, D. Atar, S. Hohnloser, G. Hindricks, and P. Kirchhof, "2012 focused update of the ESC Guidelines for the management of atrial fibrillation: an update of the 2010 ESC Guidelines for the management of atrial fibrillation–developed with the special contribution of the European Heart Rhythm Association," *Europace*, vol. 14, pp. 1385–1413, 2012.

[87] I. Kodama, K. Kamiya, and J. Toyama, "Amiodarone: ionic and cellular mechanisms of action of the most promising class III agent," *Am. J. Cardiol.*, vol. 84, pp. 20r–28r, 1999.

[88] R. Sato, S. Koumi, D. H. Singer, I. Hisatome, H. Jia, S. Eager, and J. A. Wasserstrom, "Amiodarone blocks the inward rectifier potassium channel in isolated guinea pig ventricular cells," *J. Pharmacol. Exp. Ther.*, vol. 269, pp. 1213–1219, 1994.

[89] N. Ikeda, K. Nademanee, R. Kannan, and B. N. Singh, "Electrophysiologic effects of amiodarone: experimental and clinical observation relative to serum and tissue drug concentrations," *Am. Heart J.*, vol. 108, pp. 890–898, 1984.

[90] D. P. Zipes, E. N. Prystowsky, and J. J. Heger, "Amiodarone: electrophysiologic actions, pharmacokinetics and clinical effects," *J. Am. Coll. Cardiol.*, vol. 3, pp. 1059–1071, 1984.

[91] I. Kodama, K. Kamiya, and J. Toyama, "Cellular electropharmacology of amiodarone," *Cardiovasc. Res.*, vol. 35, pp. 13–29, 1997.

[92] L. A. Siddoway, "Amiodarone: guidelines for use and monitoring," *Am. Fam. Physician*, vol. 68, pp. 2189–2196, 2003.

[93] N. Gassanov, M. Dietlein, E. Caglayan, E. Erdmann, and F. Er, "Amiodarone-induced thyroid gland dysfunctions," *Dtsch. Med. Wochenschr.*, vol. 135, pp. 807–811, 2010.

[94] C. Rombach, "Simulation der Auswirkungen von Amiodaron und Cisaprid auf das EKG mit detailierten elektrophysiologischen Modellen," Diploma Thesis, Institute of Biomedical Engineering, Universität Karlsruhe (TH), Karlsruhe, 2009.

[95] N. Butte, B. W. Bottiger, and P. Teschendorf, "Amiodaron for treatment of perioperative cardiac arrythmia : A broad spectrum antiarrythmetic agent?," *Anaesthesist*, pp. 1–10, 2008.

[96] A. Burashnikov, J. Di Diego, S. Sicouri, M. Ferreiro, L. Carlsson, and C. Antzelevitch, "Atrial-selective effects of chronic amiodarone in the management of atrial fibrillation," *Heart Rhythm*, vol. 5, pp. 1735–1742, 2008.

[97] N. Lalevee, J. Nargeot, S. Barrere-Lemaire, P. Gautier, and S. Richard, "Effects of amiodarone and dronedarone on voltage-dependent sodium current in human cardiomyocytes," *J. Cardiovasc. Electrophysiol.*, vol. 14, pp. 885–890, 2003.

[98] P. Gautier, E. Guillemare, A. Marion, J.-P. Bertrand, Y. Tourneur, and D. Nisato, "Electrophysiologic characterization of dronedarone in guinea pig ventricular cells," *J. Cardiovasc. Pharmacol.*, vol. 41, pp. 191–202, 2003.

[99] P. Gautier, E. Guillemare, L. Djandjighian, A. Marion, J. Planchenault, C. Bernhart, J.-M. Herbert, and D. Nisato, "In vivo and in vitro characterization of the novel antiarrhythmic agent SSR149744C: electrophysiological, anti-adrenergic, and anti-angiotensin II effects," *J. Cardiovasc. Pharmacol.*, vol. 44, pp. 244–257, 2004.

[100] R. Bogdan, H. Goegelein, and H. Ruetten, "Effect of dronedarone on Na^+, Ca^{2+} and HCN channels," *Naunyn-Schmied Arch. Pharmacol.*, pp. 347–356, 2011.

[101] R. L. Page, B. Hamad, and P. Kirkpatrick, "Dronedarone," *Nat. Rev. Drug Discov.*, vol. 8, pp. 769–770, 2009.

[102] A. Adlan and G. Lip, "Benefit-risk assessment of dronedarone in the treatment of atrial fibrillation," *Drug Saf.*, vol. 36, pp. 93–110, 2013.

[103] L. Kober, C. Torp-Pedersen, J. McMurray, O. Gotzsche, S. Levy, H. Crijns, J. Amlie, and J. Carlsen, "Increased mortality after dronedarone therapy for severe heart failure," *N. Engl. J. Med.*, vol. 358, pp. 2678–2687, 2008.

[104] N. Penugonda, A. Mohamand-Borowski, and J. F. Burke, "Dronedarone for atrial fibrillation: How does it compare with amiodarone?," *Clev. Clin. J. Med.*, vol. 78, pp. 179–185, 2011.

[105] Sanofi-aventis Canada Inc., Laval, Quebec H7L 4A8, *PRODUCT MONOGRAPH MULTAQ, Dronedarone Tablets*, March 2011.

[106] W. Sun, J. S. M. Sarma, and B. N. Singh, "Chronic and acute effects of dronedarone on the action potential of rabbit atrial muscle preparations: comparison with amiodarone," *J. Cardiovasc. Pharmacol.*, vol. 39, pp. 677–684, 2002.

[107] B. D. Walker, C. B. Singleton, J. A. Bursill, K. R. Wyse, S. M. Valenzuela, M. R. Qiu, S. N. Breit, and T. J. Campbell, "Inhibition of the human ether-a-go-go-related gene (HERG) potassium channel by cisapride: affinity for open and inactivated states," *B. J. Pharmacol.*, vol. 128, pp. 444–450, 1999.

[108] L. R. Wiseman and D. Faulds, "Cisapride. An updated review of its pharmacology and therapeutic efficacy as a prokinetic agent in gastrointestinal motility disorders," *Drugs*, vol. 47, pp. 116–152, 1994.

[109] D. K. Wysowski, A. Corken, H. Gallo-Torres, L. Talarico, and E. M. Rodriguez, "Postmarketing reports of QT prolongation and ventricular arrhythmia in association with cisapride and Food and Drug Administration regulatory actions," *Am. J. Gastroenterol.*, vol. 96, pp. 1698–1703, 2001.

[110] D. Rampe, M. L. Roy, A. Dennis, and A. M. Brown, "A mechanism for the proarrhythmic effects of cisapride (Propulsid): high affinity blockade of the human cardiac potassium channel HERG," *FEBS Letters*, vol. 417, pp. 28–32, 1997.

[111] C.-E. Chiang, T.-M. Wang, and H.-N. Luk, "Inhibition of L-type Ca(2+) current in Guinea pig ventricular myocytes by cisapride," *J. Biomed. Sci.*, vol. 11, pp. 303–314, 2004.

[112] L. Carlsson, G. Amos, B. Andersson, L. Drews, G. Duker, and G. Wadstedt, "Electrophysiological characterization of the prokinetic agents cisapride and mosapride in vivo and in vitro: implications for proarrhythmic potential?," *J. Pharmacol. Exp. Ther.*, vol. 282, pp. 220–227, 1997.

[113] J. Di Diego, L. Belardinelli, and C. Antzelevitch, "Cisapride-induced transmural dispersion of repolarization and torsade de pointes in the canine left ventricular wedge preparation during epicardial stimulation," *Circ.*, vol. 108, pp. 1027–1033, 2003.

[114] G. Seemann, *Modeling of electrophysiology and tension development in the human heart.* PhD thesis, Universitätsverlag Karlsruhe, Institut für Biomedizinische Technik, 2005.

[115] M. Courtemanche, R. J. Ramirez, and S. Nattel, "Ionic mechanisms underlying human atrial action potential properties: Insights from a mathematical model," *Am. J. Physiol. Heart. Circ. Physiol.*, vol. 275, pp. H301–H321, 1998.

[116] A. Nygren, C. Fiset, L. Firek, J. W. Clark, D. S. Lindblad, R. B. Clark, and W. R. Giles, "Mathematical model of a adult human atrial cell. The role of K+ currents in repolarization," *Circ. Res.*, vol. 82, pp. 63–81, 1998.

[117] M. M. Maleckar, J. L. Greenstein, N. A. Trayanova, and W. R. Giles, "Mathematical simulations of ligand-gated and cell-type specific effects on the action potential of human atrium," *Prog. Biophys. Mol. Biol.*, vol. 98, pp. 161–170, 2008.

[118] J. T. Koivumäki, T. Korhonen, and P. Tavi, "Impact of sarcoplasmic reticulum calcium release on calcium dynamics and action potential morphology in human atrial myocytes: a computational study," *PLoS Comput. Biol.*, vol. 7, p. e1001067, 2011.

[119] E. Grandi, S. V. Pandit, N. Voigt, A. J. Workman, D. Dobrev, J. Jalife, and D. M. Bers, "Human Atrial Action Potential and Ca2+ Model: Sinus Rhythm and Chronic Atrial Fibrillation," *Circ. Res.*, vol. 109, no. 9, pp. 1055–1066, 2011.

[120] O. Dössel, M. W. Krueger, F. M. Weber, M. Wilhelms, and G. Seemann, "Computational modeling of the human atrial anatomy and electrophysiology," *Med. Biol. Eng. Comput.*, vol. 50, pp. 773–799, 2012.

[121] C. H. Luo and Y. Rudy, "A dynamic model of the cardiac ventricular action potential. I. Simulations of ionic currents and concentraion changes," *Circ. Res.*, vol. 74, pp. 1071–1096, 1994.

[122] D. S. Lindblad, C. R. Murphey, J. W. Clark, and W. R. Giles, "A model of the action potential and underlying membrane currents in a rabbit atrial cell," *Am. J. Physiol. Heart. Circ. Physiol.*, vol. 271, pp. 1666–1696, 1996.

[123] J. Kneller, R. Zou, E. J. Vigmond, Z. Wang, L. J. Leon, and S. Nattel, "Cholinergic atrial fibrillation in a computer model of a two-dimensional sheet of canine atrial cells with realistic ionic properties," *Circ. Res.*, vol. 90, pp. 1–15, 2002.

[124] K. Zorn-Pauly, P. Schaffer, B. Pelzmann, P. Lang, H. Machler, B. Rigler, and B. Koidl, "If in left human atrium: a potential contributor to atrial ectopy," *Cardiovasc. Res.*, vol. 64, pp. 250–259, 2004.

[125] E. Grandi, F. S. Pasqualini, and D. M. Bers, "A novel computational model of the human ventricular action potential and Ca transient," *J. Mol. Cell. Cardiol.*, vol. 48, pp. 112–121, 2010.

[126] T. R. Shannon, F. Wang, J. Puglisi, C. Weber, and D. M. Bers, "A mathematical treatment of integrated Ca dynamics within the ventricular myocyte," *Biophys. J.*, vol. 87, pp. 3351–3371, 2004.

[127] K. ten Tusscher and A. Panfilov, "Alternans and spiral breakup in a human ventricular tissue model," *Am. J. Physiol. Heart Circ. Physiol.*, vol. 291, pp. H1088–100, 2006.

[128] K. H. W. J. ten Tusscher, D. Noble, P. J. Noble, and A. V. Panfilov, "A model for human ventricular tissue," *Am. J. Physiol. Heart. Circ. Physiol.*, vol. 286, pp. H1573–H1589, 2004.

[129] M. D. Stern, L. S. Song, H. Cheng, J. S. Sham, H. T. Yang, K. R. Boheler, and E. Ríos, "Local control models of cardiac excitation-contraction coupling. A possible role for allosteric interactions between ryanodine receptors," *J. Gen. Physiol.*, vol. 113, pp. 469–489, 1999.

[130] M. Courtemanche, R. J. Ramirez, and S. Nattel, "Ionic targets for drug therapy and atrial fibrillation-induced electrical remodeling: insights from a mathematical model," *Cardiovasc. Res.*, vol. 42, pp. 477–489, 1999.

[131] E. M. Cherry, H. M. Hastings, and S. J. Evans, "Dynamics of human atrial cell models: Restitution, memory, and intracellular calcium dynamics in single cells," *Prog. Biophys. Mol. Biol.*, vol. 98, pp. 24–37, Sept. 2008.

[132] K. Tsujimae, S. Murakami, and Y. Kurachi, "In silico study on the effects of IKur block kinetics on prolongation of human action potential after atrial fibrillation-induced electrical remodeling," *Am. J. Physiol. Heart. Circ. Physiol.*, vol. 294, pp. H793–800, 2008.

[133] H. Zhang, C. J. Garratt, J. Zhu, and A. V. Holden, "Role of up-regulation of IK1 in action potential shortening associated with atrial fibrillation in humans," *Cardiovasc. Res.*, vol. 66, pp. 493–502, 2005.

[134] N. H. L. Kuijpers, M. Potse, P. M. van Dam, H. M. M. ten Eikelder, S. Verheule, F. W. Prinzen, and U. Schotten, "Mechanoelectrical coupling enhances initiation and affects perpetuation of atrial fibrillation during acute atrial dilation," *Heart Rhythm*, vol. 8, pp. 429–436, 2011.

[135] V. Jacquemet, L. Kappenberger, and C. S. Henriquez, "Modeling atrial arrhythmias: Impact on clinical diagnosis and therapies," *IEEE Rev. Biomed. Eng.*, vol. 1, pp. 94–114, 2008.

[136] G. Seemann, D. L. Weiss, E. P. Scholz, F. B. Sachse, and O. Dössel, "Effects of a KCNA5 Mutation in a Human Atrial Myocyte: A Computational Study," in *Biophys. J (Annual Meeting Abstracts)*, 2008.

[137] S. Kharche, C. Garratt, M. Boyett, S. Inada, A. Holden, J. Hancox, and H. Zhang, "Atrial proarrhythmia due to increased inward rectifier current (I(K1)) arising from KCNJ2 mutation–a simulation study," *Prog. Biophys. Mol. Biol.*, vol. 98, pp. 186–197, 2008.

[138] A. Michailova, W. Lorentz, and A. McCulloch, "Modeling transmural heterogeneity of K(ATP) current in rabbit ventricular myocytes," *Am. J. Physiol. Cell Physiol.*, vol. 293, pp. C542–C557, 2007.

[139] J. L. Puglisi and D. M. Bers, "LabHEART: An Interactive Computer Model of Rabbit Ventricular Myocyte Ion Channels and Ca transport," *Am. J. Physiol. Cell Physiol.*, vol. 281, pp. C2049–C2060, 2001.

[140] J. Terkildsen, E. Crampin, and N. Smith, "The balance between inactivation and activation of the Na^+-K^+ pump underlies the triphasic accumulation of extracellular K^+ during myocardial ischemia," *Am. J. Physiol. Heart. Circ. Physiol.*, vol. 293, pp. H3036–45, 2007.

[141] A. Pollard, W. Cascio, V. Fast, and S. Knisley, "Modulation of triggered activity by uncoupling in the ischemic border. A model study with phase 1b-like conditions," *Cardiovasc. Res.*, vol. 56, pp. 381–392, 2002.

[142] X. Jie and N. Trayanova, "Mechanisms for initiation of reentry in acute regional ischemia phase 1B," *Heart Rhythm*, vol. 7, pp. 379–386, 2010.

[143] E. Ramirez, J. Saiz, B. Trenor, J. Ferrero, G. Molto, and V. Hernandez, "Influence of 1B ischemic ventricular tissue on the automaticity of Purkinje fibers: A simulation study," in *Computers in Cardiology*, pp. 617–620, 2007.

[144] G. M. Faber and Y. Rudy, "Action potential and contractility changes in [Na+]i overloaded cardiac myocytes: a simulation study," *Biophys. J.*, vol. 78, pp. 2392–2404, 2000.

[145] F. O. Campos, A. J. Prassl, G. Seemann, R. Weber Dos Santos, G. Plank, and E. Hofer, "Influence of ischemic core muscle fibers on surface depolarization potentials in superfused cardiac tissue preparations: a simulation study," *Med. Biol. Eng. Comput.*, vol. 50, no. 5, pp. 461–472, 2012.

[146] A. Mahajan, Y. Shiferaw, D. Sato, A. Baher, R. Olcese, L.-H. Xie, M.-J. Yang, P.-S. Chen, J. G. Restrepo, A. Karma, A. Garfinkel, Z. Qu, and J. N. Weiss, "A rabbit ventricular action potential model replicating cardiac dynamics at rapid heart rates," *Biophys. J.*, vol. 94, pp. 392–410, 2008.

[147] D. Nickerson and M. Buist, "Practical application of CellML 1.1: The integration of new mechanisms into a human ventricular myocyte model," *Prog. Biophys. Mol. Biol.*, vol. 98, pp. 38–51, 2008.

[148] D. Weiss, M. Ifland, F. B. Sachse, G. Seemann, and O. Dössel, "Modeling of cardiac ischemia in human myocytes and tissue including spatiotemporal electrophysiological variations," *Biomed. Tech.*, vol. 54, pp. 107–125, 2009.

[149] J. M. J. Ferrero, J. Saiz, J. M. Ferrero, and N. V. Thakor, "Simulation of action potentials from metabolically impaired cardiac myocytes. Role of ATP-sensitive K+ current," *Circ. Res.*, vol. 79, pp. 208–221, 1996.

[150] J. Saiz, J. Gomis-Tena, M. Monserrat, J. Ferrero, K. Cardona, and J. Chorro, "Effects of the antiarrhythmic drug dofetilide on transmural dispersion of repolarization in ventriculum. A computer modeling study," *IEEE Trans. Biomed. Eng.*, vol. 58, pp. 43–53, 2011.

[151] N. Zemzemi, M. Bernabeu, J. Saiz, and B. Rodriguez, "Simulating drug-induced effects on the heart: from ion channel to body surface electrocardiogram," in *Functional Imaging and Modeling of the Heart* (D. N. Metaxas and L. Axel, eds.), vol. 6666 of *LNCS*, pp. 259–266, Springer, 2011.

[152] B. Trenor, J. M. J. Ferrero, B. Rodriguez, and F. Montilla, "Effects of pinacidil on reentrant arrhythmias generated during acute regional ischemia: a simulation study," *Ann. Biomed. Eng.*, vol. 33, pp. 897–906, 2005.

[153] A. P. Benson, O. V. Aslanidi, H. Zhang, and A. V. Holden, "The canine virtual ventricular wall: a platform for dissecting pharmacological effects on propagation and arrhythmogenesis," *Prog. Biophys. Mol. Biol.*, vol. 96, pp. 187–208, 2008.

[154] T. J. Hund and Y. Rudy, "Rate dependence and regulation of action potential and calcium transient in a canine cardiac ventricular cell model," *Circ.*, vol. 110, pp. 3168–3174, 2004.

[155] O. V. Aslanidi, M. Al-Owais, A. P. Benson, M. Colman, C. J. Garratt, S. H. Gilbert, J. P. Greenwood, A. V. Holden, S. Kharche, E. Kinnell, E. Pervolaraki, S. Plein, J. Stott, and H. Zhang, "Virtual tissue engineering of the human atrium: modelling pharmacological actions on atrial arrhythmogenesis," *Eur. J. Pharm. Sci.*, vol. 46, pp. 209–221, Jul 2012.

[156] B. Rodriguez, K. Burrage, D. Gavaghan, V. Grau, P. Kohl, and D. Noble, "The systems biology approach to drug development: application to toxicity assessment of cardiac drugs," *Clin. Pharmacol. Ther.*, vol. 88, pp. 130–134, 2010.

[157] M. Mollova, K. Bersell, S. Walsh, J. Savla, L. Das, S.-Y. Park, L. Silberstein, C. Dos Remedios, D. Graham, S. Colan, and B. Kuhn, "Cardiomyocyte proliferation contributes to heart growth in young humans," *Proc. Natl. Acad. Sci.*, vol. 110, pp. 1446–1451, 2013.

[158] L. Tung, *A bi-domain model for describing ischemic myocardial dc potentials.* PhD thesis, Massachusetts Institute of Technology, 1978.

[159] W. T. Miller and D. B. Geselowitz, "Simulation studies of the electrocardiogram. I. The normal heart," *Circ. Res.*, vol. 43, no. 2, pp. 301–315, 1978.

[160] C. S. Henriquez, "Simulating the electrical behavior of cardiac tissue using the bidomain model," *Crit. Rev. Biomed. Eng.*, vol. 21, pp. 1–77, 1993.

[161] N. Hooke, C. S. Henriquez, P. Lanzkron, and D. Rose, "Linear algebraic transformations of the bidomain equations: Implications for numerical methods," *Math. Biosci.*, pp. 127–145, 1994.

[162] E. J. Vigmond, R. Weber dos Santos, A. J. Prassl, M. Deo, and G. Plank, "Solvers for the cardiac bidomain equations," *Prog. Biophys. Mol. Biol.*, vol. 96, pp. 3–18, 2008.

[163] D. Keller, *Multiscale modeling of the ventricles: From cellular electrophysiology to body surface electrocardiograms.* PhD thesis, Karlsruhe, KIT Scientific Publishing, 2011.

[164] G. X. Yan, W. Shimizu, and C. Antzelevitch, "The characteristics and distribution of M cells in arterially perfused canine left ventricular wedge preparations," *Circ.*, vol. 98, pp. 1921–1927, 1998.

[165] N. Szentadrassy, T. Banyasz, T. Biro, G. Szabo, B. I. Toth, J. Magyar, J. Lazar, A. Varro, L. Kovacs, and P. P. Nanasi, "Apico-basal inhomogeneity in distribution of ion channels in canine and human ventricular myocardium," *Cardiovasc. Res.*, vol. 65, pp. 851–860, 2005.

[166] D. Farina, *Forward and inverse problems of elctrocardiography: Clinical investigations.* PhD thesis, Karlsruhe, Universitätsverlag Karlsruhe, 2008.

[167] M. Fink and D. Noble, "Markov models for ion channels: versatility versus identifiability and speed," *Phil. Trans. Roy. Soc. A*, vol. 367, pp. 2161–2179, 2009.

[168] C. T. Kelley, *Iterative methods for optimization*, vol. 18, ch. 3.3 Trust Region Methods. Society for Industrial Mathematics, 1999.

[169] J. Kennedy and Eberhart, "Particle swarm optimization," in *Proc. of the IEEE Int. Conference on Neural Networks*, pp. 1942–1948, 1995.

[170] R. Poli, J. Kennedy, and T. Blackwell, "Particle swarm optimization," *Swarm Intell.*, vol. 1, pp. 33–57, 2007.

[171] M. Clerc and J. Kennedy, "The particle swarm - explosion, stability, and convergence in a multidimensional complex space," *IEEE Trans. Evol. Comput.*, vol. 6, no. 1, pp. 58–73, 2002.

[172] D. Szekely, J. Vandenberg, S. Dokos, and A. Hill, "An improved curvilinear gradient method for parameter optimization in complex biological models," *Med. Biol. Eng. Comput.*, vol. 49, pp. 289–296, 2011.

[173] J. Abbruzzese, F. B. Sachse, M. Tristani-Firouzi, and M. C. Sanguinetti, "Modification of hERG1 channel gating by Cd2+," *J. Gen. Physiol.*, vol. 136, pp. 203–224, 2010.

[174] A. Saltelli, S. Tarantola, F. Campolongo, and M. Ratto, *Sensitivity analysis in practice: a guide to assessing scientific models.* Chichester, UK: John Wiley & Sons, Ltd, 2004.

[175] G. Seemann, P. Carrillo Bustamante, S. Ponto, M. Wilhelms, E. P. Scholz, and O. Dössel, "Atrial Fibrillation-based Electrical Remodel

ing in a Computer Model of the Human Atrium," in *Comput. Cardiol.*, vol. 37, pp. 417–420, 2010.

[176] Y. Yang, J. Li, X. Lin, K. Hong, L. Wang, J. Liu, L. Li, D. Yan, D. Liang, J. Xiao, H. Jin, J. Wu, Y. Zhang, and Y.-H. Chen, "Novel KCNA5 loss-of-function mutations responsible for atrial fibrillation," *J. Hum. Genet.*, vol. 54, pp. 277–283, 2009.

[177] M.-A. Bray, S.-F. Lin, R. R. Aliev, B. J. Roth, and J. P. Wikswo, "Experimental and theoretical analysis of phase singularity dynamics in cardiac tissue," *J. Cardiovasc. Electrophysiol.*, vol. 12, pp. 716–722, 2001.

[178] M.-A. Bray and J. P. Wikswo, "Use of topological charge to determine filament location and dynamics in a numerical model of scroll wave activity," *IEEE Trans. Biomed. Eng.*, vol. 49, pp. 1086–1093, 2002.

[179] L. J. Rantner, L. Wieser, M. C. Stuhlinger, F. Hintringer, B. Tilg, and G. Fischer, "Detection of phase singularities in triangular meshes," *Method. Inf. Med.*, vol. 46, pp. 646–654, 2007.

[180] R. Zou, J. Kneller, L. J. Leon, and S. Nattel, "Development of a computer algorithm for the detection of phase singularities and initial application to analyze simulations of atrial fibrillation," *Chaos*, vol. 12, pp. 764–778, 2002.

[181] J. Ng and J. J. Goldberger, "Understanding and interpreting dominant frequency analysis of AF electrograms," *J. Cardiovasc. Electrophysiol.*, vol. 18, pp. 680–685, 2007.

[182] R. Plonsey and R. Barr, *Bioelectricity: a quantitative approach*, ch. Sources and Fields. New York: Springer, 3 ed., 2007.

[183] R. M. Shaw and Y. Rudy, "The vulnerable window for unidirectional block in cardiac tissue: Characterization and dependence on membrane excitability and intercellular coupling," *J. Cardiovasc. Electrophysiol.*, vol. 6, pp. 115–131, 1995.

[184] B. Trenor, L. Romero, J. M. J. Ferrero, J. Saiz, G. Molto, and J. M. Alonso, "Vulnerability to reentry in a regionally ischemic tissue: a simulation study," *Ann. Biomed. Eng.*, vol. 35, pp. 1756–1770, 2007.

[185] J. N. Weiss, A. Garfinkel, H. S. Karagueuzian, Z. Qu, and P. S. Chen, "Chaos and the transition to ventricular fibrillation: a new approach to antiarrhythmic drug evaluation," *Circ.*, vol. 99, pp. 2819–2826, 1999.

[186] A. Garfinkel, Y. H. Kim, O. Voroshilovsky, Z. Qu, J. R. Kil, M. H. Lee, H. S. Karagueuzian, J. N. Weiss, and P. S. Chen, "Preventing ventricular fibrillation by flattening cardiac restitution," *Proc. Natl. Acad. Sci.*, vol. 97, pp. 6061–6066, 2000.

[187] W. Haverkamp, G. Breithardt, A. J. Camm, M. J. Janse, M. R. Rosen, C. Antzelevitch, D. Escande, M. Franz, M. Malik, A. Moss, and R. Shah, "The potential for QT prolongation and pro-arrhythmia by non-anti-arrhythmic drugs: clinical and regulatory implications. Report on a Policy Conference of the European Society of Cardiology," *Cardiovasc. Res.*, vol. 47, pp. 219–233, 2000.

[188] Z. Syed, E. Vigmond, S. Nattel, and L. J. Leon, "Atrial cell action potential parameter fitting using genetic algorithms," *Med. Biol. Eng. Comput.*, vol. 43, pp. 561–571, 2005.

[189] B. Hui, S. Dokos, and N. Lovell, "Parameter identifiability of cardiac ionic models using a novel CellML least squares optimization tool," in *Engineering in Medicine and Biology Society, 2007. EMBS 2007. 29th Annual International Conference of the IEEE*, pp. 5307–5310, 2007.

[190] A. Bueno-Orovio, E. M. Cherry, and F. H. Fenton, "Minimal model for human ventricular action potentials in tissue," *J. Theor. Biol.*, vol. 253, pp. 544–560, 2008.

[191] F. Chen, A. Chu, X. Yang, Y. Lei, and J. Chu, "Identification of the parameters of the Beeler-Reuter ionic equation with a partially perturbed particle swarm optimization," *IEEE Trans. Biomed. Eng.*, vol. 59, pp. 3412–3421, 2012.

[192] G. Seemann, S. Lurz, D. U. J. Keller, D. L. Weiss, E. P. Scholz, and O. Dössel, "Adaption of Mathematical Ion Channel Models to measured data using the Particle Swarm Optimization," in *IFMBE Proceedings*, vol. 22, pp. 2507–2510, 2008.

[193] K. Hong, P. Bjerregaard, I. Gussak, and R. Brugada, "Short QT syndrome and atrial fibrillation caused by mutation in KCNH2," *J. Cardiovasc. Electrophysiol.*, vol. 16, pp. 394–396, 2005.

[194] M. Sinner, A. Pfeufer, M. Akyol, B.-M. Beckmann, M. Hinterseer, A. Wacker, S. Perz, W. Sauter, T. Illig, M. Nabauer, C. Schmitt, H.-E. Wichmann, A. Schomig, G. Steinbeck, T. Meitinger, and S. Kaab, "The non-synonymous coding IKr-channel variant KCNH2-K897T

is associated with atrial fibrillation: results from a systematic candidate gene-based analysis of KCNH2 (HERG)," *Eur Heart. J.*, vol. 29, pp. 907–914, 2008.

[195] A. van Oosterom and V. Jacquemet, "Ensuring stability in models of atrial kinetics," in *Computers in Cardiology*, pp. 69–72, 2009.

[196] A. J. Workman, K. A. Kane, and A. C. Rankin, "The Contribution of Ionic Currents to Changes in Refractoriness of Human Atrial Myocytes Associated with chronic Atrial Fibrillation," *Cardiovasc. Res.*, vol. 52, pp. 226–235, 2001.

[197] T. Christ, E. Wettwer, N. Voigt, O. Hala, S. Radicke, K. Matschke, A. Varro, D. Dobrev, and U. Ravens, "Pathology-specific effects of the IKur/Ito/IK,ACh blocker AVE0118 on ion channels in human chronic atrial fibrillation," *Br. J. Pharmacol.*, vol. 154, no. 8, pp. 1619–1630, 2008.

[198] M. R. Franz, "Electrical remodeling of the human atrium: similar effects in patients with chronic atrial fibrillation and atrial flutter," *J. Am. Coll. Cardiol.*, vol. 30, pp. 1785–1792, 1997.

[199] W. C. Yu, S. H. Lee, and C. T. Tai, "Reversal of atrial electrical remodeling following cardioversion of long-standing atrial fibrillation in man," *Cardiovasc. Res.*, vol. 42, pp. 470–476, 1999.

[200] G. K. Feld, M. Mollerus, U. Birgersdotter-Green, O. Fujimura, T. D. Bahnson, K. Boyce, and M. Rahme, "Conduction velocity in the tricuspid valve-inferior vena cava isthmus is slower in patients with type I atrial flutter compared to those without a history of atrial flutter," *J. Cardiovasc. Electrophysiol.*, vol. 8, pp. 1338–1348, 1997.

[201] A. S. Agha, C. A. Castillo, A. J. Castellanos, R. J. Myerburg, and M. P. Tessler, "Supernormal conduction in the human atria," *Circ.*, vol. 46, pp. 522–527, 1972.

[202] A. Nygren, L. J. Leon, and W. R. Giles, "Simulations of the human atrial action potential," *Phil. Trans. R. Soc. Lond. A*, vol. 359, pp. 1111–1125, 2001.

[203] E. M. Cherry and S. J. Evans, "Properties of two human atrial cell models in tissue: Restitution, memory, propagation, and reentry," *J. Theor. Biol.*, vol. 254, pp. 674–690, Oct. 2008.

[204] F. M. Weber, *Personalizing simulations of the human atria: Intracardiac measurements, tissue conductivities, and cellular electro-*

physiology. PhD thesis, Karlsruhe, KIT Scientific Publishing, 2011.

[205] T. M. Olson, A. E. Alekseev, X. K. Liu, S. Park, L. V. Zingman, M. Bienengraeber, S. Sattiraju, J. D. Ballew, A. Jahangir, and A. Terzic, "Kv1.5 channelopathy due to KCNA5 loss-of-function mutation causes human atrial fibrillation," *Hum. Mol. Genet.*, vol. 15, pp. 2185–2191, 2006.

[206] E. Wettwer, O. Hala, T. Christ, J. F. Heubach, D. Dobrev, M. Knaut, A. Varró, and U. Ravens, "Role of IKur in controlling action potential shape and contractility in the human atrium: influence of chronic atrial fibrillation," *Circ.*, vol. 110, pp. 2299–2306, 2004.

[207] M. Potse, B. Dubé, J. Richter, A. Vinet, and R. Gulrajani, "A Comparison of Monodomain and Bidomain Reaction-Diffusion Models for Action Potential Propagation in the Human Heart," *IEEE Trans. Biomed. Eng.*, vol. 53, pp. 2425–2435, 2006.

[208] M. Wilhelms, "Analysis of cardiac ischemia regarding ECG and stability in a computer model of the human ventricles," Diploma Thesis, Institute of Biomedical Engineering, Universität Karlsruhe (TH), Karlsruhe, 2009.

[209] J. Ornato, I. Menown, M. Peberdy, M. Kontos, J. Riddell, G. r. Higgins, S. Maynard, and J. Adgey, "Body surface mapping vs 12-lead electrocardiography to detect ST-elevation myocardial infarction," *Am. J. Emerg. Med.*, vol. 27, pp. 779–784, 2009.

[210] A. Pashaei, C. Hoogendoorn, R. Sebastián, D. Romero, O. Cámara, and A. F. Frangi, "Effect of Scar Development on Fast Electrophysiological Models of the Human Heart: In-Silico Study on Atlas-Based Virtual Populations," in *Functional Imaging and Modeling of the Heart* (D. N. Metaxas and L. Axel, eds.), vol. 6666 of *LNCS*, pp. 427–436, Springer, 2011.

[211] L. X. Cubeddu, "QT prolongation and fatal arrhythmias: a review of clinical implications and effects of drugs," *Am. J. Ther.*, vol. 10, pp. 452–457, 2003.

[212] J. Kneller, J. Kalifa, R. Zou, A. V. Zaitsev, M. Warren, O. Berenfeld, E. J. Vigmond, L. J. Leon, S. Nattel, and J. Jalife, "Mechanisms of atrial fibrillation termination by pure sodium channel blockade in an ionically-realistic mathematical model," *Circ. Res.*, vol. 96, pp. e35–47, 2005.

[213] C. Antzelevitch, "Ionic, molecular, and cellular bases of QT-interval prolongation and torsade de pointes," *Europace*, vol. 9 Suppl 4, pp. iv4–15, 2007.

[214] B. Darpö, "Detection and reporting of drug-induced proarrhythmias: room for improvement," *Europace*, vol. 9 Suppl 4, pp. iv23–36, 2007.

[215] E. Drouin, G. Lande, and F. Charpentier, "Amiodarone reduces transmural heterogeneity of repolarization in the human heart," *J. Am. Coll. Cardiol.*, vol. 32, pp. 1063–1067, 1998.

[216] F. Morady, L. DiCarlo, R. Krol, J. Baerman, and M. de Buitleir, "Acute and chronic effects of amiodarone on ventricular refractoriness, intraventricular conduction and ventricular tachycardia induction," *J. Am. Coll. Cardiol.*, vol. 7, pp. 148–157, 1986.

[217] C. Omichi, S. Zhou, M.-H. Lee, A. Naik, C.-M. Chang, A. Garfinkel, J. Weiss, S.-F. Lin, H. Karagueuzian, and P.-S. Chen, "Effects of amiodarone on wave front dynamics during ventricular fibrillation in isolated swine right ventricle," *Am. J. Physiol. Heart Circ. Physiol.*, vol. 282, pp. H1063–70, 2002.

Karlsruhe Transactions on Biomedical Engineering
(ISSN 1864-5933)

Karlsruhe Institute of Technology / Institute of Biomedical Engineering (Ed.)

Die Bände sind unter www.ksp.kit.edu als PDF frei verfügbar oder als Druckausgabe bestellbar.

Band 2   Matthias Reumann
Computer assisted optimisation on non-pharmacological treatment of congestive heart failure and supraventricular arrhythmia. 2007
ISBN 978-3-86644-122-4

Band 3   Antoun Khawaja
Automatic ECG analysis using principal component analysis and wavelet transformation. 2007
ISBN 978-3-86644-132-3

Band 4   Dmytro Farina
Forward and inverse problems of electrocardiography: clinical investigations. 2008
ISBN 978-3-86644-219-1

Band 5   Jörn Thiele
Optische und mechanische Messungen von elektrophysiologischen Vorgängen im Myokardgewebe. 2008
ISBN 978-3-86644-240-5

Band 6   Raz Miri
Computer assisted optimization of cardiac resynchronization therapy. 2009
ISBN 978-3-86644-360-0

Band 7   Frank Kreuder
2D-3D-Registrierung mit Parameterentkopplung für die Patientenlagerung in der Strahlentherapie. 2009
ISBN 978-3-86644-376-1

Band 8   Daniel Unholtz
Optische Oberflächensignalmessung mit Mikrolinsen-Detektoren für die Kleintierbildgebung. 2009
ISBN 978-3-86644-423-2

Band 9   Yuan Jiang
Solving the inverse problem of electrocardiography in a realistic environment. 2010
ISBN 978-3-86644-486-7

Karlsruhe Transactions on Biomedical Engineering
(ISSN 1864-5933)

Karlsruhe Transactions on Biomedical Engineering
(ISSN 1864-5933)

Band 19   Martin Wolfgang Krüger
Personalized Multi-Scale Modeling of the Atria: Heterogeneities, Fiber Architecture, Hemodialysis and Ablation Therapy. 2012
ISBN 978-3-86644-948-0

Band 20   Mathias Wilhelms
Multiscale Modeling of Cardiac Electrophysiology: Adaptation to Atrial and Ventricular Rhythm Disorders and Pharmacological Treatment. 2013
ISBN 978-3-7315-0045-2